Essentials of Biology

Essentials of Biology

Jerry Thorpe

Larsen & Keller
www.larsen-keller.com

Essentials of Biology
Jerry Thorpe
ISBN: 978-1-64172-129-5 (Hardback)

Larsen & Keller
Published by Larsen and Keller Education,
5 Penn Plaza,
19th Floor,
New York, NY 10001, USA

Cataloging-in-Publication Data

Essentials of biology / Jerry Thorpe.
p. cm.
Includes bibliographical references and index.
ISBN 978-1-64172-129-5
1. Biology. 2. Life sciences. I. Thorpe, Jerry.
QH313 .E87 2019
574--dc23

The publisher's policy is to use permanent paper from mills that operate a sustainable forestry policy. Furthermore, the publisher ensures that the text paper and cover boards used have met acceptable environmental accreditation standards.

Printed and bound in China.

For more information regarding Larsen and Keller Education and its products, please visit the publisher's website www.larsen-keller.com

Table of Contents

Preface

Biology is a natural science that is concerned with the study of living organisms and the organization of life at chemical, molecular, physical and physiological levels. Cell theory, genetics, homeostasis, evolution and energy provide the theoretical foundations of biology.

Biology is a vast discipline that differentiates into a number of significant sub-disciplines such as anatomy, biochemistry, biological engineering, biophysics, botany, zoology, developmental biology, ecology, evolutionary biology, genetics, marine biology, etc. This book is compiled in such a manner, that it will provide in-depth knowledge about the principles and applications of biology. Such selected concepts that redefine this discipline have been presented herein. This book is an essential guide for both academicians and those who wish to pursue this discipline further.

A detailed account of the significant topics covered in this book is provided below:

Chapter 1, Biology is a natural science concerned with the study of life and living organisms. The investigation of the physical structure, molecular interactions, chemical processes, physiological mechanisms and evolution is under the scope of this field. This chapter has been carefully written to provide an introduction to the varied principles of biology for an easy understanding. **Chapter 2**, The cell is the fundamental functional, structural and biological unit of life. It is of two types, eukaryotic and prokaryotic cells. The latter are seen in unicellular organisms while former maybe either unicellular or multicellular. This chapter closely examines the different kinds of cells and cellular organizations, as well as the vital concepts of cell structure, cell theory, cellular processes and function of cell. **Chapter 3**, Genetics is concerned with the study of genes, variation and heredity in organisms. To develop an understanding of biology, it is vital to delve into the fundamentals of DNA, RNA, chromosomes, genetic inheritance, genetic drift as well as the process of meiosis and mitosis, which have been extensively covered in this chapter. **Chapter 4**, Any change expressed in the hereditable traits of organisms over successive generations is termed as evolution. Such changes are mediated by genes when they are inherited from parent to offspring during the process of reproduction. The aim of this chapter is to provide an insight into the phenomenon of evolution through the elucidation of topics such as genetic variation, mutation and Darwin's theory. **Chapter 5**, Reproduction is a natural process that occurs in living organisms by which new individuals are created. Reproduction is of two types, sexual and asexual. The topics elaborated in this chapter on plant reproduction, animal reproduction and human reproduction system will help in providing a comprehensive understanding of the processes of reproduction. **Chapter 6**, Population ecology is a sub-discipline of ecology. The study of the interaction of species population with the environment and the change in population sizes over time and space is under the scope of population ecology. All the vital topics of this field like population density, population age structure, population growth, population dispersion and limits to population growth have been carefully examined in this chapter. **Chapter 7**, Biology is a vast subject that branches out into significant sub-disciplines, which makes it essential to understand its various sub-domains. This chapter discusses these branches in extensive detail such as botany, zoology, marine biology, molecular biology, etc. for a holistic understanding of the subject.

It gives me an immense pleasure to thank our entire team for their efforts. Finally in the end, I would like to thank my family and colleagues who have been a great source of inspiration and support.

Jerry Thorpe

1

Introduction to Biology

Biology is a natural science concerned with the study of life and living organisms. The investigation of the physical structure, molecular interactions, chemical processes, physiological mechanisms and evolution is under the scope of this field. This chapter has been carefully written to provide an introduction to the varied principles of biology for an easy understanding.

Biology is the "science of life." It is the study of living and once-living things, from submicroscopic structures in single-celled organisms to entire ecosystems with billions of interacting organisms; it further ranges in time focus from a single metabolic reaction inside a cell to the life history of one individual and on to the course of many species over eons of time. Biologists study the characteristics and behaviors of organisms, how species and individuals come into existence, and their interactions with each other and with the environment. The purview of biology extends from the origin of life to the fundamental nature of human beings and their relationship to all other forms of life.

Biology, or "life science," offers a window into fundamental principles shared by living organisms. These principles reveal a harmony and unity of the living world operating simultaneously among a great diversity of species and even in the midst of competition both between and within species for scarce resources. The overlying harmony is seen at each level, from within a cell to the level of systems in individuals (nervous, circulatory, respiratory, etc.), the immediate interactions of one organism with others, and on to the complex of organisms and interactions comprising an ecosystem with a multitude of ecological niches each supporting one species. Such harmony is manifested in many universally shared characteristics among living beings, including interdependence, a common carbon-based biochemistry, a widespread pattern of complementary polarities, sexual reproduction, and homeostasis.

As the science dealing with all life, biology encompasses a broad spectrum of academic fields that have often been viewed as independent disciplines. Among these are molecular biology, biochemistry, cell biology, physiology, anatomy, developmental biology, genetics, ecology, paleontology, and evolutionary biology. While competition among individuals expressing genetic variability has generally been identified as a key factor in evolutionary development, the pivotal roles of cooperation and long-term symbiosis or symbiogenesis in living systems have emerged in the late twentieth century as essential complementary focal points for understanding both the origin of species and the dynamics of biological systems.

The word "biology" derives from Greek and is generally rendered as "study of life." The word "biology" in its modern sense seems to have been introduced independently by Gottfried Reinhold Treviranus and by Jean-Baptiste Lamarck.

Principles of Biology

While biology is unlike physics in that it does not usually describe biological systems in terms of objects that exclusively obey immutable physical laws described by mathematics, it is nevertheless characterized by several major principles and concepts, which include: universality, evolution, interactions, diversity, and continuity.

Universality: Cells, Biochemistry, Energy, Development, Homeostasis, and Polarity

Living organisms share many universal characteristics, including that they are composed of cells; pass on their heredity using a nearly universal genetic code; need energy from the environment to exist, grow, and reproduce; maintain their internal environment; and exhibit dual characteristics or complementary polarities. This are the common set of characteristics identified by biologists that distinguish living organisms from nonliving things.

With the exception of viruses, all organisms consist of cells, which are the basic units of life, being the smallest unit that can carry on all the processes of life, including maintenance, growth, and even self-repair. Some simple life forms, such as the paramecium, consist of a single cell throughout their life cycle and are called unicellular organisms. Multicellular organisms, such as a whale or tree, may have trillions of cells differentiated into many diverse types each performing a specific function.

All cells, in turn, are based on a carbon-based biochemistry, and all organisms pass on their heredity via genetic material based on nucleic acids such as DNA using a nearly universal genetic code. Every cell, no matter how simple or complex, utilizes nucleic acids for transmitting and storing the information needed for manufacturing proteins.

Every living being needs energy from the environment in order to exist, grow, and reproduce. Radiation from the sun is the main source of energy for life and is captured through photosynthesis, the biochemical process in which plants, algae, and some bacteria harness the energy of sunlight to produce food. Ultimately, nearly all living things depend on energy produced from photosynthesis for their nourishment, making it vital to life on Earth. There are also some bacteria that utilize the oxidation of inorganic compounds such as hydrogen sulfide or ferrous iron as an energy source. An organism that produces organic compounds from carbon dioxide as a carbon source, using either light or reactions of inorganic chemical compounds as a source of energy, is called an autotroph. Other organisms do not make their own food but depend directly or indirectly on autotrophs for their food. These are called heterotrophs.

In development, the theme of universal processes is also present. Living things grow and develop as they age. In most metazoan organisms the basic steps of the early embryo development share similar morphological stages and include similar genes.

All living organisms, whether unicellular or multicellular, exhibit homeostasis. Homeostasis is the property of an open system to regulate its internal environment so as to maintain a stable condition. Homeostasis can manifest itself at the cellular level through the maintenance of a stable internal acidity (pH); at the organismal level, warm-blooded animals maintain a constant internal body temperature; and at the level of the ecosystem, for example when atmospheric carbon dioxide levels rise, plants are theoretically able to grow healthier and thus remove more carbon dioxide from the atmosphere. Tissues and organs can also maintain homeostasis.

In addition, living beings share with all existent beings the quality of dual characteristics or complementary polarities. One common pair of dual characteristics is the quality of positivity and negativity: Just as sub-atomic particles have positive (electron) and negative (proton) elements that interrelate and form atoms, living beings commonly exhibit positive and negative characteristics. Most animals reproduce through relationships between male and female, and higher plants likewise have male and female elements, such as the (male) stamen and (female) pistil in flowering plants (angiosperms). Lower plants, fungi, some of the protists, and bacteria likewise exhibit reproductive variances, which are usually symbolized by + and - signs (rather than being called male and female), and referred to as "mating strains" or "reproductive types" or similar appellations.

Another more philosophical concept is the universal dual characteristic of within each organism of the invisible, internal character or nature and the visible aspects of matter, structure, and shape. For example, an animal will exhibit the internal aspects of life, instinct, and function of its cells, tissues, and organs, which relate with the visible shape made up by those cells, tissues, and organs.

Sexual reproduction is a trait that is almost universal among eukaryotes. Asexual reproduction is not uncommon among living organisms. In fact, it is widespread among fungi and bacteria, many insects reproduce in this manner, and some reptiles and amphibians. Nonetheless, with the exception of bacteria (prokaryotes), sexual reproduction is also seen in these same groups. (Some treat the unidirectional lateral transfer of genetic material in bacteria, between donors (+ mating type) and recipients (- mating type), as a type of sexual reproduction.) Evolutionary biologist and geneticist John Maynard Smith maintained that the perceived advantage for an individual organism to pass only its own entire genome to its offspring is so great that there must be an advantage by at least a factor of two to explain why nearly all animal species maintain a male sex.

Another characteristic of living things is that they take substances from the environment and organize them in complex hierarchical levels. For example, in multicellular organisms, cells are organized into tissues, tissues are organized into organs, and organs are organized into systems.

In addition, all living beings respond to the environment; that is, they react to a stimulus. A cockroach may respond to light by running for a dark place. When there is a complex set of response, it is called a behavior. For example, the migration of salmon is a behavioral response.

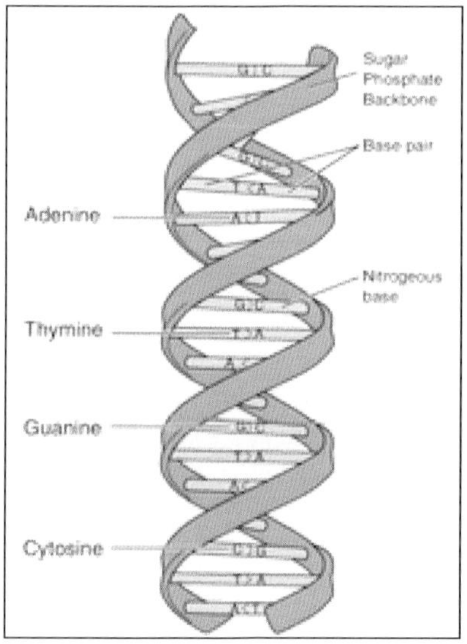

Schematic representation of DNA the primary genetic material

Evolution: A Common Organizing Principle of Biology

A central, organizing concept in biology is that all life has descended from a common origin through a process of evolution. Indeed, eminent evolutionist Theodosius Dobzhansky has stated that "Nothing in biology makes sense except in the light of evolution." Evolution can be considered a unifying theme of biology because the concept of descent with modification helps to explain the common carbon-based biochemistry, the nearly universal genetic code, and the similarities and relationships among living organisms, as well as between organisms of the past with organisms today.

Evolutionary theory actually comprises several distinct components. Two of the major strands are the theory of descent with modification, which addresses the "pattern" of evolution, and the theory of natural selection, which addresses the "process" of evolution. Charles Darwin established evolution as a viable theory by marshaling and systematizing considerable evidence for the theory of descent with modification, including evidence from paleontology, classification, biogeography, morphology, and embryology. The mechanism that Darwin postulated, natural selection, aims to account for evolutionary changes at both the microevolutionary level (i.e., gene changes on the populational level) and the macroevolutionary level (i.e., major transitions between species and origination of new designs). Experimental tests and observations provide strong evidence for microevolutionary change directed by natural selection operating on heritable expressed variation, while evidence that natural selection directs macroevolution is limited to fossil evidence of some key transition sequences and extrapolation from evidences on the microevolutionary level.

The evolutionary history of a species—which tells the characteristics of the various species from which it descended—together with its genealogical relationship to every other species is called its phylogeny. Widely varied approaches to biology generate information about phylogeny. These include the comparisons of DNA sequences conducted within molecular biology or genomics, and comparisons of fossils or other records of ancient organisms in paleontology. Biologists organize and analyze evolutionary relationships through various methods, including phylogenetics, phenetics, and cladistics. Major events in the evolution of life, as biologists currently understand them, are summarized on an evolutionary timeline.

Interactions: Harmony and Bi-level Functionality

Symbiosis between clownfish of the genus Amphiprion that dwell among the tentacles of tropical sea anemones. The territorial fish protects the anemone from anemone-eating fish, and in turn the stinging tentacles of the anemone protect the clownfish from its predators.

Every living thing interacts with other organisms and its environment. One of the reasons that biological systems can be difficult to study is that there are so many different possible interactions with other organisms and the environment. A microscopic bacterium responding to a local gradient in sugar is as much responding to its environment as a lion is responding to its environment when it is searching for food in the African savanna. Within a particular species, behaviors can be cooperative, aggressive, parasitic, or symbiotic.

Matters become more complex still when two or more different species interact in an ecosystem, studies of which lie in the province of ecology. Analysis of ecosystems shows that a major factor in maintaining harmony and reducing competition is the tendency for each species to find and occupy a distinctive niche not occupied by other species.

Overlying the interactions of organisms is a sense of unity and harmony at each level of interaction. On the global level, for example, one can see the harmony between plant and animal life in terms of photosynthesis and respiration. Plants, through photosynthesis, use carbon dioxide and give off oxygen. While they also respire, plants' net input to the globe is considerably more oxygen than they consume (with algae in the ocean being a major source of planetary oxygen). Animals, on the other hand, consume oxygen and discharge carbon dioxide.

On the trophic level, the food web demonstrates harmony. Plants convert and store the sun's energy. These plants serve as food for herbivores, which in turn serve as food for carnivores, which are consumed by top carnivores. Top carnivores (and species at all other trophic levels), when dead, are broken down by decomposers such as bacteria, fungi, and some insects into minerals and humus in the soil, which is then used by plants.

On the level of individuals, the remarkable harmony among systems (nervous, circulatory, respiratory, endocrine, reproductive, skeletal, digestive, etc.) is a wonder to behold. Even within a cell, one sees remarkable examples of unity and harmony, such as when a cell provides a product to the body (such as a hormone) and receives oxygen and nourishment from the body. So remarkable is the harmony evident among organisms, and between organisms and the environment, that some have proposed a theory that the entire globe acts as if one, giant, functioning organism (the Gaia theory). According to well-known biologist Lynn Margulis and science writer Dorion Sagan, even evolution is tied to cooperation and mutual dependence among organisms: "Life did not take over the globe by combat, but by networking."

An underlying explanation for such observed harmony is the concept of bi-level functionality, the view that every entity exists in an integral relation with other entities in ways that permit an individual entity to advance its own multiplication, development, self-preservation, and self-strengthening (a function for the individual) while at the same time contribute toward maintaining or developing the larger whole (a function for the whole). These functions are not independent but interdependent. The individual's own success allows it to contribute to the whole, and while the individual contributes something of value to the larger entity, assisting the larger entity in advancing its own function, the larger entity likewise provides the environment for the success of the individual.

For example, in the cells of a multicellular organism, each cell provides a useful function for the body as a whole. A cell's function may be to convert sugar to ADP energy, attack foreign invaders, or produce hormones. A cell in the epithelial tissue of the stomach may secrete the enzyme pepsin to help with digestion. The cell's function of providing pepsin to the body is harmonized with the body's needs for maintenance, development, and reproduction. The body, on the other hand, supports the individual cell and its function by providing food, oxygen, and other necessary materials, and by transporting away the toxic waste materials. Each cell actually depends on the other cells in the body to perform their functions and thus keep the body in proper functioning order. Likewise, a particular taxonomic group (taxa) not only advances its own survival and reproduction, but also provides a function for the ecosystems of which it is part, such as the ocelot species helping to regulate prey populations and thus help ecosystems to maintain balance. An ecosystem provides an environment for the success of this taxonomic group and thus its contribution to the ecosystem. In essence, this explanation holds that while animals and plants may seem to struggle against one another for existence, in reality they do not. Rather, they all contribute to the whole, in harmony.

Human beings, the most complex of all biological organisms, likewise live in a biosphere that is all interrelated and is necessary for physical life. Thus, it becomes essential that human beings, as the most powerful of all life forms and in many ways an encapsulation of the whole (a "microcosm of creation" according to a theological perspective), understand and care for the environment. In religious terms, this is sometimes referred to as the "third blessing," the role of humankind to love and care for creation. The science of biology is central to this process.

The science of physics offers complementary rationales both for explaining evolutionary development and also for urging humans to love and care for the biosphere. This striking advance in physics arises through the extension of the second law of thermodynamics to apply to "open" systems, which include all forms of life. The extended second law states simply that natural processes in open systems tend to dissipate order as rapidly as possible. From this perspective, evolution of life's successively more ordered and complex systems occurs because the greater a system's order and complexity, the greater its capacity to dissipate order. Human beings, as the planet's dominant and most complex species, face a thermodynamic imperative to apply themselves toward establishing an even greater level of order and dynamic complexity on the planet. Achieving such greater order would likely require that humans learn to live together in peace while living in synergy with the biosphere.

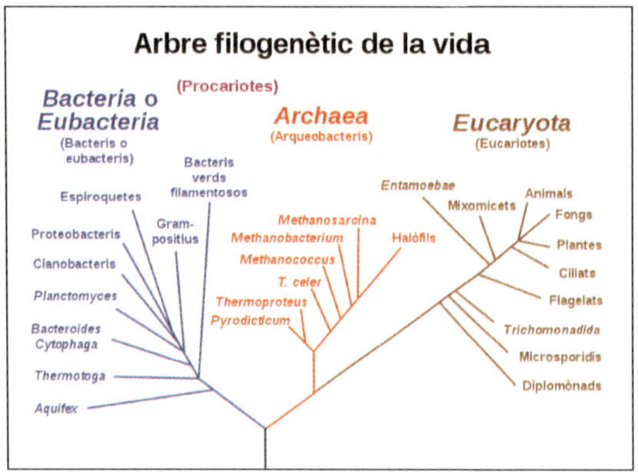

A phylogenetic tree of all living things, based on rRNA gene data, showing the separation of the three domains bacteria, archaea, and eukaryotes as described initially by Carl Woese. Trees constructed with other genes are generally similar, although they may place some early branching groups very differently. The exact relationships of the three domains are still being debated.

Diversity: The Variety of Living Organisms

Despite the underlying unity, life exhibits an astonishing wide diversity in morphology, behavior, and life histories. In order to grapple with this diversity, biologists, following a conventional western scientific approach and historically unaware of the profound interdependence of all life on the planet, attempt to classify all living things. This scientific classification should reflect the evolutionary trees (phylogenetic trees) of the different organisms. Such classifications are the province of the disciplines of systematics and taxonomy. Taxonomy puts organisms in groups called taxa, while systematics seeks their relationships.

Until the nineteenth century, living organisms were generally divided into two kingdoms: animal and plant, or the Animalia and the Plantae. As evidence accumulated that these divisions were insufficient to express the diversity of life, schemes with three, four, or more kingdoms were proposed.

A popular scheme, developed in 1969 by Robert Whitaker, delineates living organisms into five kingdoms:

Monera — Protista — Fungi — Plantae —Animalia.

In the six-kingdom classification, the six top-level groupings (kingdoms) are:

Archaebacteria, Monera (the bacteria and cyanobacteria), Protista, Fungi, Plantae, and Animalia.

These schemes coexist with another scheme that divides living organisms into the two main divisions of prokaryote (cells that lack a nucleus: bacteria, etc.) and eukaryote (cells that have a nucleus and membrane-bound organelles: animals, plants, fungi, and protists).

In 1990, another scheme, a three-domain system, was introduced by Carl Woese and has became very popular (with the "domain" a classification level higher than kingdom):

Archaea (originally Archaebacteria) — Bacteria (originally Eubacteria) — Eukaryota (or Eucarya).

The three-domain system is a biological classification that emphasizes his separation of prokaryotes into two groups, the Bacteria and the Archaea (originally called Eubacteria and Archaebacteria). When recent work revealed that what were once called "prokaryotes" are far more diverse than suspected, the prokaryotes were divided into the two domains of the Bacteria and the Archaea, which are considered to be as different from each other as either is from the eukaryotes. Woese argued based on differences in 16S ribosomal RNA genes that these two groups and the eukaryotes each arose separately from an ancestral progenote with poorly developed genetic machinery. To reflect these primary lines of descent, he treated each as a domain, divided into several different kingdoms. The groups were also renamed the Bacteria, Archaea, and Eukaryota, further emphasizing the separate identity of the two prokaryote groups.

There is also a series of intracellular "parasites" that are progressively less alive in terms of being metabolically active:

Viruses — Viroids — Prions

Continuity: The Common Descent of Life

A group of organisms is said to have common descent if they have a common ancestor. All existing organisms on Earth are descended from a common ancestor or ancestral gene pool. This "last universal common ancestor," that is, the most recent common ancestor of all organisms, is believed to have appeared about 3.5 billion years ago.

The notion that "all life [is] from [an] egg" is a foundational concept of modern biology, it means that there has been an unbroken continuity of life from the initial origin of life to the present time. Up into the nineteenth century it was commonly believed that life forms can appear spontaneously under certain conditions (abiogenesis).

The universality of the genetic code is generally regarded by biologists as strong support of the theory of universal common descent (UCD) for all bacteria, archaea, and eukaryotes.

Scope of Biology

Academic Disciplines

Biologists study life over a wide range of scales: Life is studied at the atomic and molecular scale in molecular biology, biochemistry, and molecular genetics. At the level of the cell, life is studied in cell biology, and at multicellular scales, it is examined in physiology, anatomy, and histology. Developmental biology involves study of life at the level of the development or ontogeny of an individual organism.

Moving up the scale toward more than one organism, genetics considers how heredity works between parent and offspring. Ethology considers group behavior of organisms. Population genetics looks at the level of an entire population, and systematics considers the multi-species scale of lineages. Interdependent populations and their habitats are examined in ecology.

Two broad disciplines within biology are botany, the study of plants, and zoology, the study of animals. Paleontology is inquiry into the developing history of life on earth, based on working with fossils, and includes the main subfields of paleobotany, paleozoology, and micropaleontology. Changes over time, whether within populations (microevolution) or involving either speciation or the introduction of major designs (macroevolution), is part of the field of inquiry of evolutionary biology. A speculative new field is astrobiology (or xenobiology) which examines the possibility of life beyond the Earth.

Biology has become such a vast research enterprise that it is not generally studied as a single discipline, but as a number of clustered sub-disciplines. Four broad groupings are considered here. The first broad group consists of disciplines that study the basic structures of living systems: cells, genes, and so forth; a second grouping considers the operation of these structures at the level of tissues, organs and bodies; a third grouping considers organisms and their histories; and a final constellation of disciplines focuses on the interactions. It is important to note, however, that these groupings are a simplified description of biological research. In reality, the boundaries between disciplines are very fluid and most disciplines borrow techniques from each other frequently. For example, evolutionary biology leans heavily on techniques from molecular biology to determine DNA sequences that assist in understanding the genetic variation of a population; and physiology borrows extensively from cell biology in describing the function of organ systems.

Ethical Aspects

As in all sciences, biological disciplines are best pursued by persons committed to high ethical standards, maintaining the highest integrity and following a good research methodology. Data should be interpreted honestly, and results that do not fit one's preconceived biases should not be discarded or ignored in favor of data that fits one's prejudices. A biologist who puts her or his own well-being first (money, popularity, position, etc.), runs the risk of faulty or even fraudulent research. But even well-meaning biologists have gone off course in trying to fit research findings to personal biases.

Also overlying work in many biological fields is the more specific concept of bioethics. This is the discipline dealing with the ethical implications of biological research and its applications. Aspects of biology raising issues of bioethics include cloning, genetic engineering, population control, medical research on animals, creation of biological weapons, and so forth.

Structure of Life

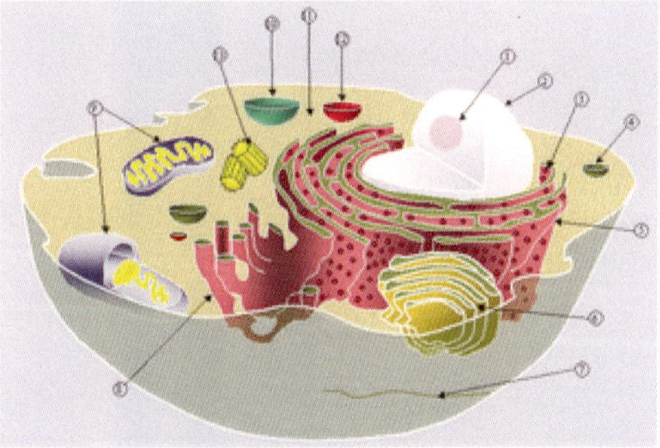

Schematic of typical animal cell depicting the various organelles and structures

Molecular biology is the study of biology at the molecular level. The field overlaps with other areas of biology, particularly genetics and biochemistry. Molecular biology chiefly concerns itself with understanding the interactions between the various systems of a cell, especially by mapping the interactions between DNA, RNA, and protein synthesis and learning how these interactions are regulated.

Cell biology studies the physiological properties of cells, as well as their behaviors, interactions, and environment; this is done both on a microscopic and molecular level. Cell biology researches both single-celled organisms like bacteria and specialized cells in multicellular organisms like humans.

Understanding the composition of cells and how cells work is fundamental to all of the biological sciences. Appreciating the similarities and differences between cell types is particularly important to the fields of cell and molecular biology. These fundamental similarities and differences provide a unifying theme, allowing the principles learned from studying one cell type to be extrapolated and generalized to other cell types.

Genetics is the science of genes, heredity, and the variation of organisms. In modern research, genetics provides important tools in the investigation of the function of a particular gene (e.g., analysis of genetic interactions). Within organisms, genetic information generally is carried in chromosomes, where it is represented in the chemical structure of particular DNA molecules.

Genes encode the information necessary for synthesizing proteins, which in turn play a large role in influencing the final phenotype of the organism, although in many instances do not completely determine it.

Developmental biology studies the process by which organisms grow and develop. Originating in embryology, today, developmental biology studies the genetic control of cell growth, differentiation, and "morphogenesis," which is the process that gives rise to tissues, organs, and anatomy. Model organisms for developmental biology include the round worm Caenorhabditis elegans, the fruit fly Drosophila melanogaster, the zebrafish Brachydanio rerio, the mouse Mus musculus, and the small flowering mustard plant Arabidopsis thaliana.

Physiology of Organisms

Physiology studies the mechanical, physical, and biochemical processes of living organisms, by attempting to understand how all the structures function as a whole. The theme of "structure to function" is central to biology.

Physiological studies have traditionally been divided into plant physiology and animal physiology, but the principles of physiology are universal, regardless of the particular organism being studied. For example, what is learned about the physiology of yeast cells can also apply to other cells. The field of animal physiology extends the tools and methods of human physiology to non-human animal species. Plant physiology also borrows techniques from both fields.

Anatomy is an important part of physiology and considers how organ systems in animals such as the nervous, immune, endocrine, respiratory, and circulatory systems function and interact. The study of these systems is shared with the medically oriented disciplines of neurology, immunology, and the like. The field of health science deals with both human and animal health.

Diversity and Evolution of Organisms

Evolutionary biology is concerned with the origin and descent of species, and their change over time, i.e., their evolution. Evolutionary biology is an inclusive field because it includes scientists from many traditional taxonomically oriented disciplines. For example, it generally includes scientists who may have a specialist training in particular organisms such as mammalogy, ornithology, or herpetology, but uses those organisms as systems to answer general questions in evolution. It also generally includes paleontologists who use fossils to answer questions about the mode and tempo of evolution, as well as theoreticians in areas such as population genetics and evolutionary theory. In the 1990s, developmental biology made a re-entry into evolutionary biology from its initial exclusion from the modern synthesis through the study of evolutionary developmental biology. Related fields which are often considered part of evolutionary biology are phylogenetics, systematics, and taxonomy.

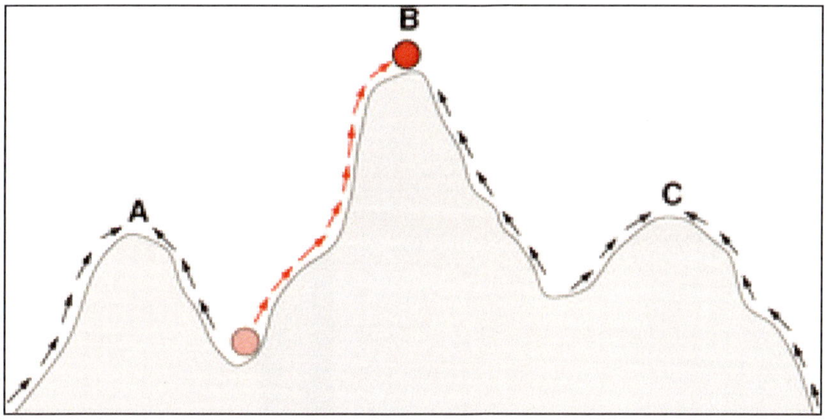

In population genetics the evolution of a population of organisms is sometimes depicted as if traveling on a fitness landscape. The arrows indicate the preferred flow of a population on the landscape, and the points A, B, and C are local optima. The red ball indicates a population that moves from a very low fitness value to the top of a peak.

The two major traditional taxonomically oriented disciplines are botany and zoology. Botany is the scientific study of plants. It covers a wide range of scientific disciplines that study the growth, reproduction, metabolism, development, diseases, and evolution of plant life. Zoology is the discipline that involves the study of animals, which includes the physiology of animals studied under various fields, including anatomy and embryology. The common genetic and developmental mechanisms of animals and plants is studied in molecular biology, molecular genetics, and developmental biology. The ecology of animals is covered under behavioral ecology and other fields.

Classification of Life

The dominant classification system is called Linnaean taxonomy, which includes ranks and binomial nomenclature. How organisms are named is governed by international agreements such as the International Code of Botanical Nomenclature (ICBN), the International Code of Zoological Nomenclature (ICZN), and the International Code of Nomenclature of Bacteria (ICNB). A fourth Draft BioCode was published in 1997 in an attempt to standardize naming in the three areas, but it has not yet been formally adopted. The International Code of Virus Classification and Nomenclature (ICVCN) remains outside the BioCode.

Interactions of Organisms

Ecology studies the distribution and abundance of living organisms, and the interactions between organisms and their environment. The environment of an organism includes both its habitat, which can be described as the sum of local abiotic factors like climate and geology, as well as the other organisms which share its habitat. Ecological systems are studied at several different levels—from individuals and populations to ecosystems and the biosphere level. Ecology is a multi-disciplinary science, drawing on many other branches of science.

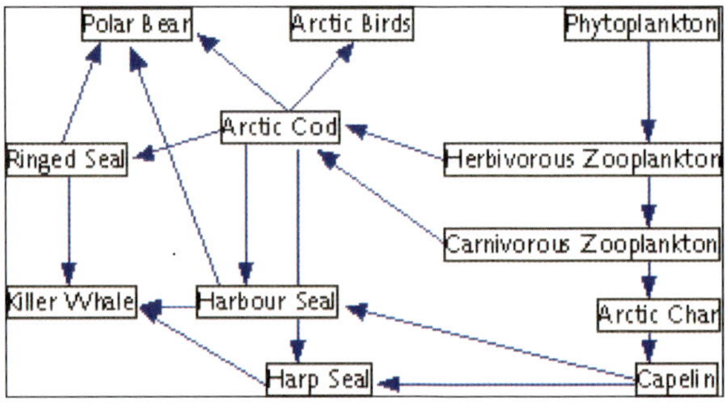

A food web, a generalization of the food chain, depicting the complex interrelationships among organisms in an ecosystem.

Ethology studies animal behavior (particularly of social animals such as primates and canids), and is sometimes considered as a branch of zoology. Ethologists have been particularly concerned with the evolution of behavior and the understanding of behavior in terms of evolutionary thought.

2

Cells

The cell is the fundamental functional, structural and biological unit of life. It is of two types, eukaryotic and prokaryotic cells. The latter are seen in unicellular organisms while former maybe either unicellular or multicellular. This chapter closely examines the different kinds of cells and cellular organizations, as well as the vital concepts of cell structure, cell theory, cellular processes and function of cell.

A cell is the basic unit of life as we know it. It is the smallest unit capable of independent reproduction. Robert Hooke suggested the name 'cell' in 1665, from the Latin cella meaning storeroom or chamber, after using a very early microscope to look at a piece of cork.

It is also said that he thought that the rectangular chambers looked like the cells in some monasteries.

Physically cells always have a boundary membrane, a little like a polythene bag encloses contents within it. Inside the space limited by the membrane there is an remarkable chemical processing unit.

From the point of view of cell structure biologists divide organisms into two groups, the bacteria (the prokaryotes), and all other animals and plants (the eucaryotes).

In bacteria chemical reactions take place almost anywhere within the cell. Bacteria contain genetic information in the form of DNA but it is not confined within a sac called a nucleus.

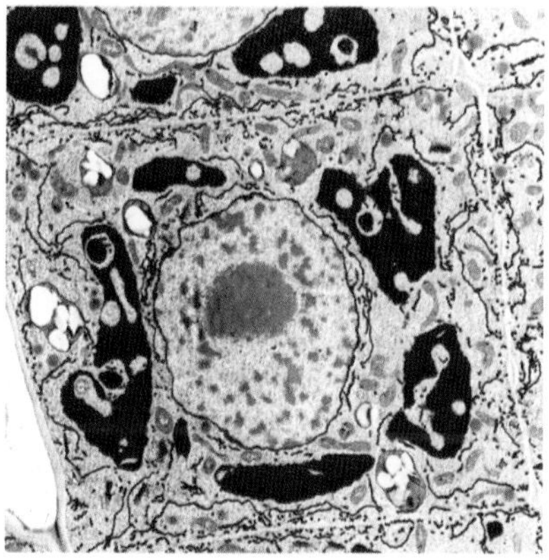

The main part of this image shows a root tip meristem cell from pea root. The black (osmium) staining shows the nuclear envelope, endoplasmic reticulum, Golgi apparatus and vacuoles.

In higher animals and plants specific functions are carried out by specialised structures. Collectively these are called organelles and include structures that contain the construction and operating plans of the cell (the nucleus), protein manufacturing areas (ribosomes), energy conversion units (mitochondria) and protein modifying and fat production areas (endoplasmic reticulum).

Additionally in plants there are light energy absorbers and converters (chloroplasts). Chloroplasts are almost unique in their capacity to convert sunlight energy into carbohydrate.

Cells also contain an elaborate transport network of filaments and fibres (the cytoskeleton) and a liquid (cytosol).

On the outer surface of a cell there can be a sticky material called extracellular matrix. This is proving to be very important to the cells it surrounds. Some animal cells produce bone and cartilage. Plant and animal cells have many features in common but plant cells also have a distinct rigid cell wall. Many plant cells also have large fluid filled sacs called vacuoles and some contain types of thickening that give plants rigidity and wood its unique strength.

Such is the efficiency of the cell that the main simple basic structure and function has been conserved during evolution and dispersal since cells started to form about 3.5 billion years ago.

The capacity and productivity of cells is truly amazing. In bacteria for example all the instructions come from a single closed loop of DNA. Each cell can divide in 20 minutes and given suitable conditions can keep dividing to produce 5 billion cells in eleven hours. Cells of this type produce some 400 different proteins and these are produced by enzyme assisted chemical reactions working at the rate of 100 times a second. This is why diseases such as meningitis and food poisoning can attack a person so quickly.

There is no such thing as a typical cell but most cells have chemical and structural features in common.

This is very important from the point of view of cell and molecular biology. It means that biologists can work on a cell from a mouse and be reasonably certain that the same processes will occur in a similar cell in a lion, a human or a fruit fly. This is possible because all cells are thought to have arisen from a common ancestor.

Many different types of plant and animal cells have evolved. In humans there are about 200 different types but within cells there only about 20 different structures or organelles.

Many cells carry out specialised functions; this is what makes them different. The specialisation of cells depends almost always on the exaggeration of properties common to cells. Cells lining the intestine for example have extended cell walls that increase the amount of surface area that is available to absorb food.

Nerve cells can be very long, extending for example in humans from the base of the spine to the foot.

Cells in heart muscle process a lot of energy and this is carried out by the high number of mitochondria found in these cells. At a molecular level however all cells resemble one another.

Cells vary greatly in their relative size although similar cells tend to be of similar size. Unfortunately most cells cannot be seen without a microscope.

The eggs of frogs and birds are large but they are made of a cell and a very large food store linked together. In relative size it has been suggested that the difference between the size of a bacterial cell and the egg of a frog would be the difference between a person and a frog's egg half a mile in diameter.

Cells are remarkable structures and in addition to facts mentioned already, they are able to communicate with each other receiving and rejecting messages.

Animal Cell

Animal cells are the basic unit of life in organisms of the kingdom Animalia. They are eukaryotic cells, meaning that they have a true nucleus and specialized structures called organelles that carry out different functions. Plants, fungi, and protists all have eukaryotic cells, while bacteria and archaea have simpler prokaryotic cells. Animal cells do not have cell walls or chloroplasts, the organelle that carries out photosynthesis.

Functions of Animal Cells

Cells carry out all the processes of the body, including producing and storing energy, and making proteins, which are molecules that have roles in metabolism, DNA replication, and transportation of other molecules. Cells are highly specialized to carry out specific tasks. For example, the heart has cardiac muscle cells that beat in unison, and digestive tract cells have cilia, finger-like projections that increase surface area for the absorption of nutrients during digestion. There are over 200 different types of cells in the human body. Red blood cells contain hemoglobin, the molecule that carries oxygen, and they have no nuclei; this is a specialization that allows each red blood cell to carry as much oxygen within it as possible.

Multiple cells form tissues, which are organized groups of cells that carry out a certain function. In turn, groups of similar tissues form the body's organs, such as the brain, lungs, and heart. Organs work together in organ systems, like the nervous system, digestive system, and circulatory system. Organ systems vary depending on the species. For example, insects have open circulatory systems, where blood is pumped directly into body cavities and surrounds their tissues Vertebrates such as fish, mammals, and birds, on the other hand, have closed circulatory systems; their blood is enclosed within blood vessels where it travels to target tissues. Cells are the "building blocks" of organ systems.

Animal Cell Structure

The cell has a variety of different parts. It contains many different types of specialized organelles that carry out all of its functions. Not every animal cell has all types of organelles, but in general, animal cells do contain most if not all of the following organelles.

Parts of an animal cell.

Nucleus

The nucleus contains a cell's deoxyribonucleic acid (DNA), its genetic material. DNA contains instructions for making proteins, which controls all of the body's activities. In the nucleus, DNA is tightly winded around histones, which are proteins, to form structures called chromosomes. The nucleus regulates which genes are expressed in the cell, which controls the cell's activity and functioning and will be different depending on the type of cell. DNA is located in the nucleolus region of the nucleus, where ribosomes are made. The nucleus is surrounded by a nuclear envelope (also called nuclear membrane), which separates it from the rest of the cell.

The nucleus also regulates the growth and division of the cell. When the cell is preparing to divide during mitosis, the chromosomes in the nucleus duplicate and separate, and two daughter cells are formed. Organelles called centrosomes help organize DNA during cell division. Cells usually have one nucleus each.

Ribosomes

Ribosomes are where proteins are synthesized. They are found within all cells, including animal cells. In the nucleus, a sequence of DNA that codes for a specific protein is copied onto a complimentary messenger RNA (mRNA) chain. The mRNA chain travels to the ribosome via transfer RNA (tRNA), and its sequence is used to determine the correct placement of amino acids in a chain that makes up the protein. In animal cells, ribosomes can be found freely in a cell's cytoplasm, or attached to membranes of the endoplasmic reticulum.

Endoplasmic Reticulum

The endoplasmic reticulum (ER) is a network of membranous sacs called cisternae that branches off from the outer nuclear membrane. It modifies and transports proteins that are made by ribosomes. There are two kinds of endoplasmic reticulum, smooth and rough. Rough ER has ribosomes attached. Smooth ER does not have ribosomes attached and has functions in making lipids and steroid hormones and removing toxic substances.

Vesicles

Vesicles are small spheres of lipid bilayer, which also makes up the cell's outer membrane. They are used for transporting molecules throughout the cell from one organelle to another, and are also involved in metabolism. Specialized vesicles called lysosomes contain enzymes that digest large molecules like carbohydrates, lipids, and proteins into smaller ones so that they can be used by the cell.

Golgi Apparatus

The Golgi apparatus, also called the Golgi complex or Golgi body, is also made up of cisternae, but the cisternae are not interconnected like those of the ER. The Golgi apparatus receives proteins from the ER and folds, sorts, and packages these proteins into vesicles.

Mitochondria

The process of cellular respiration occurs in the mitochondria. During this process, sugars and fats are broken down and energy is released in the form of adenosine triphosphate (ATP). ATP powers all cellular processes, and mitochondria produce a cell's ATP, so mitochondria are commonly known as "the powerhouse of the cell".

Cytosol

Cytosol is the liquid contained within cells. Cytosol and all the organelles within it, except for the nucleus, are collectively referred to as a cell's cytoplasm. Cytosol is mostly made of water, and also contains ions like potassium, proteins, and small molecules. The pH is neutral, around 7.

Cytoskeleton

The cytoskeleton is a network of filaments and tubules found throughout the cytoplasm of the cell. It has many functions; it gives the cell shape, provides strength, stabilizes tissues, anchors organelles within the cell, and has a role in cell signaling. There are three types of cytoskeletal filaments: microfilaments, microtubules, and intermediate filaments. Microfilaments are the smallest, while microtubules are the biggest.

Cell Membrane

The cell membrane surrounds the entire cell. Animal cells only have a cell membrane; they do not have a cell wall as plant cells do. The cell membrane is a double layer made up of phospholipids. Phospholipids are molecules with a phosphate group head attached to a glycerol and two fatty acid tails. They spontaneously form double membranes in water due to the hydrophilic properties of the head and hydrophobic properties of the tails. The cell membrane is selectively permeable, meaning it only allows certain molecules through. Oxygen and carbon dioxide pass through easily, while larger or charged molecules must go through a special channel in the membrane. This maintains homeostasis within the cell.

Plant Cell

A plant cell is defined as a cell which has a nucleus and organelles bounded by a thick and rigid cell wall. Within this rigid cell wall, there are chloroplast organelles containing chlorophyll. Chlorophyll converts sunlight into energy and gives plant cells their green color.

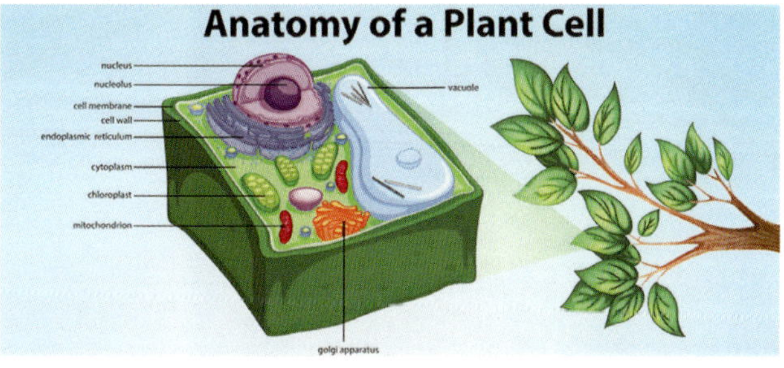

Plant Cell

Shown below is a plant cell model. Notice the nucleus and all the organelles are all contained within the rigid cell membrane. These organelles look much different when looking at a cell under a microscope. Also, start memorizing each organelle name and what they look like. This will help out a bunch, once we start looking at each organelle individually.

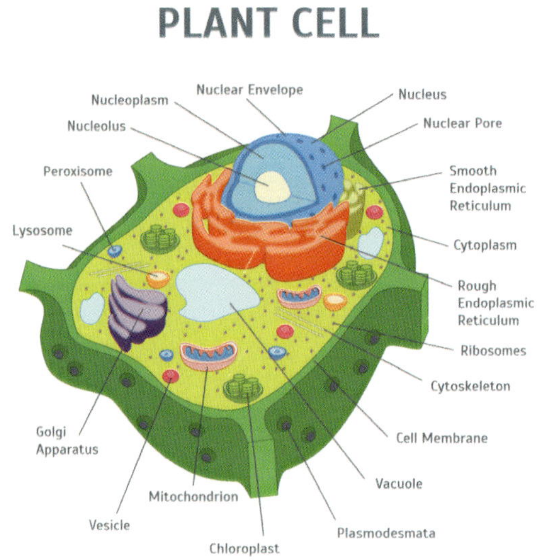

Plant cell diagram with each organelle labeled

A plant cell consists of three distinct components:

i) Cell wall,

ii) Protoplasm, and

iii) Vacuole.

The protoplasm is the living part of the cell. It is externally bounded by cell membrane or plasma membrane. The cytoplasm contains several cell organelles namely mitochondria, plastids, ribosomes, endoplasmic reticulum, lysosomes etc. (figure).

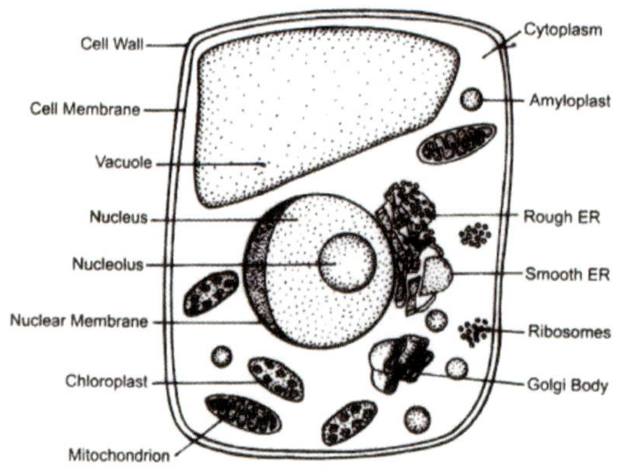

Figure: Structure of plant cell

Cell wall

Cell wall is the non-living protective layer outside the plasma membrane in the plant cells, bacteria, fungi and algae. The synthesis of cell wall in controlled by Golgi bodies. In bacteria the cell wall is composed of protein and non-cellulosic carbohydrates while in most algae, fungi and all plant cells, the cell-wall is formed of cellulose. Cell wall provides mechanical support and gives a definite shape to the cell. It protects plasma membrane and helps in imbibition's of water and movement of solutes towards protoplasm.

Protoplasm

Protoplasm is the living, colourless, elastic, colloidal semi fluid substance present in the cell. Protoplasm with non-living inclusions is called protoplast. Water is the chief constituent of an active protoplast and normally constitutes 90% of the system. The remaining parts are organic and inorganic materials.

Each protoplast keeps itself in communication with neighbouring protoplasts through small openings in the cell wall known as plasmodesmata. Protoplasm consists of cytoplasm and nucleus and is externally bounded by the cell membrane or plasmalemma.

Cell Membrane

It is a thin film like pliable membrane, and serves as protective covering of the cell. Cell membrane mainly consists of proteins and lipids but in certain cases, polysachharides have also been found. It facilitates the entrance of nutrients into the cells and allows exit of nitrogenous wastes, regulates the passage of materials into and out of the cells. It controls and maintains differential distribution of ions inside and outside the cell.

Cytoplasm

It is a jelly like fluid mass of protoplasm excluding the nucleus and surrounded by plasma membrane on the outside. It is semi permeable in nature. The cytoplasm is composed of matrix; the membrane bound organelles and non-living inclusions like vacuoles and granules. The living cytoplasmic organelles are the site of various important metabolic activities such as photosynthesis, respiration, protein synthesis etc.

Plastids

Plastids are the largest cytoplasmic organelles bounded by double membranes.

There are three types of plastids:

- Leucoplast

  These are colourless plastids found in storage organs where light is not available e.g. underground stem, roots and deeper tissues. Plastids without pigments are called leucoplasts. They store starch, fats or proteins in meristematic, embryonic and germ cells.

- Chromoplast

  Coloured plastids are called chromoplasts. These may contain red, yellow, brown, purple, blue or green pigments. These are mostly found in petals of flowers and fruits.

- Chloroplast

In a plant cell, chloroplasts are the most prominent forms of plastids that contain chlorophyll, the green pigment. The chlorophyll enables the chloroplast to harness kinetic solar energy and trap it in the form of potential energy. All living organisms directly or indirectly depend on them for energy. Chloroplast in enclosed in two smooth membranes separated by a distinct periplastidial space. The interior of chloroplast is differentiated into two parts— The Stroma and the Grana.

Stroma is the colourless ground substance that fills the chloroplast. It contains cholorophyll bearing double membranes lamellae that form flattened sac like structures called thylakoides collectively called Grana. Quantasomes are the smallest units present on the inner surface of thylakoides capable of carrying out photochemical reactions (figure below).

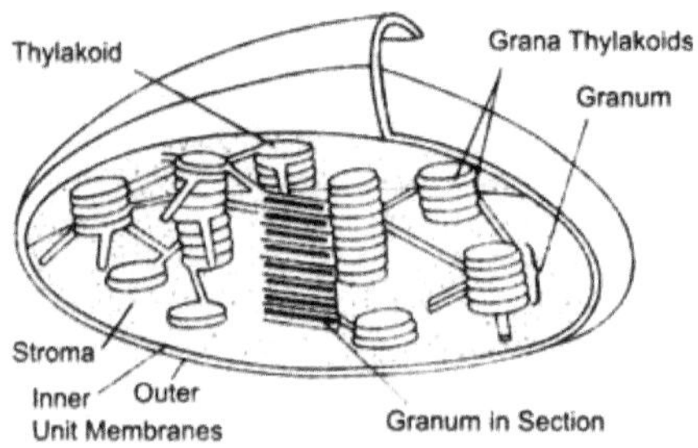

Figure: Chloroplast

Plastids, their occurrence and functions				
Kinds of Plastids		Occurrence	Functions	Pigments
(i)	Chloroplast	Higher plants and Green algae	Photosynthesis	Chlorophyll –a and Chlorophyll- b
(ii)	Phacoplast	Brown algae, diatoms and dinoflagellates	Light absorption	Fucoxanthin
(iii)	Rhodoplast	Red algae	Light absorption	Phycoerythrin
(iv)	Blue-green algae	Blue green algae	Photosynthesis	Phycocyanin, Phycocrythrin. Chlorophyll- a
(v)	Chromoplasts of Photosynthetic bacteria	Purple and non-purple sulfur bacteria	Absorption of infra-red portion of light	Bacteriochlorophyll
(vi)	Carotenoids	Tomato, red pepper, flower parts, fruits, bacteria and fungi	-	Lycopene and Capsanthin
(vii)	Amyloplasts	Storage tubers, cotyledons and endosperm	Starch storage	None
(viii)	Elaioplasts	Epidermal cells of Orchidaceae and Liliaceae	Oil storage	None
(ix)	Aleuroplasts	Seeds of Ricinus and Brazil nuts	Protein storage	None

Ribosomes

Ribosomes are the submicroscopic organelles. These are site of protein synthesis in the cell. These are found in all the cells either attached to the membranes of endoplasmic reticulum or scattered in the cytoplasm. The ribosomes are spheroidal bodies. In prokaryotic cells (Bacteria) the ribosomes are abour 15 nm and in eukaryotic cells about 25 nm.

Ribosomes from prokaryotes exist as 70S units and ribosomes in eukaryotes exust as 80S units. A ribosome is formed of two subunits – a large subunit and a small subunit. The small subunit forms a sort of cap on the flat surface of large subunit. The two units of bacterial ribosome (70S) are represented by 50S and 30S subunits, and eukaryotic ribosomes (80S) are represented by 60S and 40S subunits.

The two subunits of ribosomes usually exist free in the cytoplasm and join only during protein synthesis when a number of ribosomes get attached to mRNA in a linear fashion. These groups or clusters of ribosomes are known as Polyribosomes. The larger subunits (i.e. 60S and 50S) are attached to the membrane of endoplasmic reticulum and the smaller subunits are then bound to larger subunits. The cleft separating the two subunits lies parallel to the membrane. The messenger RNA is held by the smaller subunit, while tRNA molecule is bound to the larger subunit.

Mitochondria

Mitochondria are sausage- shaped spherical or thread like organelles present in the cytoplasm. They break down the complex carbohydrates and sugars into usable forms and supply energy for the cell, they are also called as the powerhouse of the cell.

The mitochondria are surrounded by a double walled membrane known as outer and inner membranes. The spaces between these two membranes are known as perimitochondrial space. The outer membrane is smooth but the inner membrane is variously folded into thin cristae. Inner membrane is covered with special particles called Oxysomes, these are the sites of aerobic respiration. (figure).

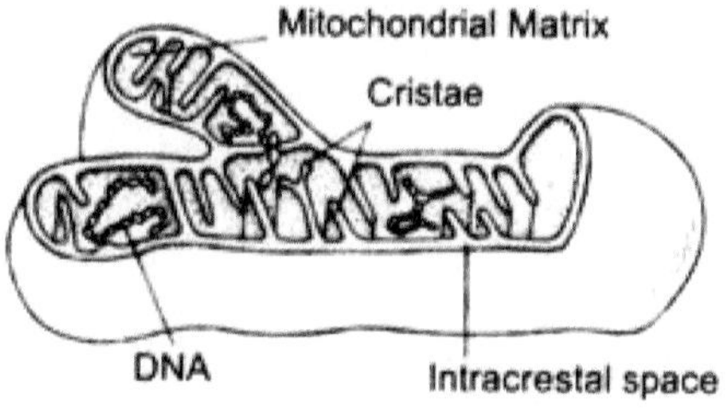

Figure: Mitochondria

Mitochondria as Semi-autonomous Organelle

Mitochondria are self perpetuating, semi-autonomous bodies. These are regarded as intracellular parasitic prokaryotes that have established symbiotic relationship with the cell.

The homologies between mitochondria and bacteria are:

- Similarity in the localization of enzymes- In bacteria the enzymes of electron transport system are localized in the plasma membrane, which can be compared to the inner membrane of mitochondria.
- Crests - I n certain bacteria the plasma membrane is projected inward into finger-like processes. These are known as mesosomes and are comparable to the mitochondrial crests. These contain respiratory chain enzymes.

- Presence of DNA- The mitochondrial DNA is circular and similar to that found in bacteria. It carries informations required for mitochondrial growth and replication.
- Presence of ribosomes- Mitochondria contains ribosomes similar to bacterial ribosomes.
- Synthesis of RNA and Proteins- Mitochondria synthesize certain amount of protein.
- Chromosome replication- Chromosome replication in mitochondria is independent of chromosomes replication in the nucleus of the cell.

Nucleus

The nucleus is the most important part of the cell which regulates all metabolic and hereditary activities within the cell It is more or less spherical, lying in the cytoplasm and occupying about two-thirds of the cell space. A typical nucleus is composed of the following structures (figure).

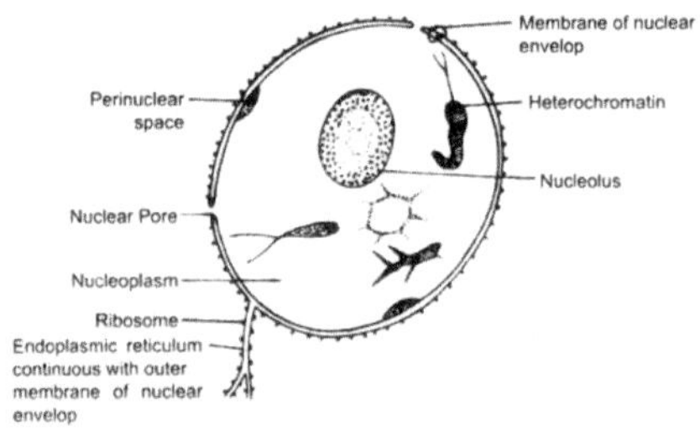

Figure: Nucleus

- Nuclear membrane

It is a selectively permeable envelope-like structure surrounding the nucleolus and nucleoplasm. It is formed of two layers separated by a fluid-filled perinuclear space. The nuclear membrane disappears during prophase stage of nuclear division and reorganizes during telophase. It regulates the passage of ions, small molecules and macromolecules of ribosomal subunits, mRNA, t RNA etc.

- Nucleoplasm

Nucleoplasm is the transparent semifluid ground substance formed of a mixture of proteins, phosphorus and some nucleic acids. Chromatin fibres or the chromonemata remain suspended in the nucleoplasm.

- Chromatin network

Chromatin fibres form a network in the nucleoplasm called chromatin net work or nuclear reticulum. Chromatin fibres are the sites of main genetic material which controls all the activities of the cell, metabolism and heredity. During cell division, the chromatin threads condense and form thick chromosomes.

- Nucleolus

The nucleolus is a spherical body lying in the nucleoplasm closely associated to the nucleolar organizer region of the chromosome. It was first described by Schleiden in 1838. It contains large amount of RNA though DNA is also present. Its main function is the synthesis of ribosomal RNA (rRNA), which helps in synthesis of ribosomes.

Golgi Body

Golgi body is also referred to as golgi complex or golgi apparatus. It plays a major role in transporting chemical substances in and out of the cell. It has three distinct components flattened sac or cisternae, clusters of transition tubules and vesicles and large vesicles or vacuoles. Golgi is mainly associated with secretory activity of the cell. It is also associated with the concentration, storage, condensation and packaging of materials for export from the cell across plasmalemma.

Endoplasmic Reticulum

Endoplasmic reticulum (ER) is the connecting link between the nucleus and cytoplasm of the plant cell. Basically, it is a network of interconnected and convoluted sacs that are located in the cytoplasm. Based on the presence or absence of ribosomes, ER can be of smooth or rough types. The former type lacks ribosomes, while the latter is covered with ribosomes. Overall, endoplasmic reticulum serves as a manufacturing, storing and transporting structure for glycogen, proteins, steroids and other compounds.

Lysosomes

Lysosomes are tiny membrane-bound, vesicular structure of cytoplasm which enclose hydrolytic enzymes and perform intracellular digestion. These are also known as suicidal bags. These are found in all animal cells but only in few plant cells.

Types of Lysosomes

Based on function or stage of digestion lysosomes are of following four types:

i) Primary lysosomes are newly formed lysosomes with hydrolytic enzymes.

ii) Secondary lysosomes are newly formed by the fusion of phagosomes and primary lysosomes. Here contents of phagosome are digested or hydrolysed.

iii) Residual bodies are exhausted secondary lysosomes. These contain undigested remains.

iv) Autophagic vacuoles are formed by the fusion of primary lysosomes with cell organelles from cell's own cytoplasm. This brings about autodigestion of cell or its organelles.

Functions

Lysosomes bring about digestion of extracellular and intracellular materials. Because of this basic character lysosomes perform following functions:

i) Lysosomes of granular leucocytes or macrophages devour foreign substances and microbes which enter the cell and guard our body against infection.

ii) Lysosomes remove worn out cell organelles, dead cell and provide energy during starvation by controlled breakdown of stored food substances.

iii) During dedifferentiation of tissues and cells, the lysosomes dissolve the specialized parts of the cells. This helps in regeneration of damaged tissue or damaged part and formation of bone from cartilage.

iv) The lytic enzymes of sperm acrosome help in the penetration of sperm into the ovum.

v) During metamorphosis, reabsorption of various larval structures like external gills of tadpole, the tail of tadpole in frog or the larval organs in the pupa of various insects is brought about by autolytic action of lysosomes.

vi) Lysosomes bring about cellular breakdown associated with ageing.

vii) Lysosomes may cause cancer by breaking down chromosomes.

Peroxisomes

Peroxisomes occur widely both in plant and animal cells. They are spherical or ovoid bodies surrounded by a single membrane. It contains certain oxidative enzymes, used for the metabolic breakdown of fatty acids into simple sugar forms. In green plants, peroxisomes help in undergoing photorespiration.

Vacuoles

Vacuoles are sap- filled vesicles in the cytoplasm. These are surrounded by a membrane called tonoplast. In a plant cell, there can be more than one vacuole; however, the centrally located vacuole is larger than others.

Tonoplast is a semi permeable membrane; it enables the vacuoles to concentrate and store nutrients and waste products. It facilitates the rapid exchange of solutes aid gases between the cytoplasm and adjoining fluids.

Cilia and Flagella

Cilia and flagella are motile hair -like appendages on the free surfaces of the cells. These are cytoplasmic processes and create water currents, food currents, act as sensory organs and perform several other functions of the cell. Cilia and flagella can be differentiated on the basis of their size, however, other physiological and morphological characteristics are almost the same.

The main difference between cilia and flagella are as follows:

Cilia	*Flagella*
5-10 long	Upto 150 long
Occur in large number	Usually one or two per cell, may be as many as eight
Cilia beat in a coordinated fashion	Flagella beat independently

Cilia and flagella are cylindrical processes projecting from the free surface of the cell. These originate from their basal bodies embedded in the cytoplasm. The basal bodies form their kinetic centres. A ciliam or flagellum consists of a longitudinal axoneme enclosed in a spiral sheath of cytoplasm and a plasma membrane continuous with the cell membrane.

Cell Structure

Cells, structurally, are units of organization. They are formally classified on the basis of where they get their energy. Prokaryotes include two of the six taxonomic kingdoms, Archaebacteria and Monera; all of these species are single-celled and most are bacteria, and they date back an astonishing 3.5 billion years or so (about 80 percent of the estimated age of the Earth itself). Eukaryotes are a "mere" 1.5 billion years old and include Animalia, Plantae, Fungae and Protista. Most eukaryotes are multicellular, although some (e.g., yeast) are not.

Prokaryotic cells, at an absolute minimum, feature an agglomeration of genetic material in the form of DNA inside an enclosure bounded by a cell membrane, also called a plasma membrane. Within this enclosure is, also, is cytoplasm, which in prokaryotes has the consistency of wet asphalt; in eukaryotes, it is much more fluid. In addition, many prokaryotes also have a cell wall outside the cell membrane to serve as a protective layer (the cell membrane serves a variety of functions). Notably, plant cells, which are eukaryotic, also include cell walls. But prokaryotic cells do not include organelles, and this is the primary structural distinction. Even if one chooses to view the distinction as a metabolic one, this is still linked to the respective structural properties.

Some prokaryotes have flagella, which are whip-like polypeptides used for propulsion. Some also have pili, which are hair-like projections used for adhesive purposes. Bacteria also come in multiple shapes: Cocci are round (like the meningococci, which can cause meningitis in humans), baccilli (rods, like the species that cause anthrax), and spirilla or spirochetes (helical bacteria, like those responsible for causing syphilis).

What about viruses? These are merely tiny bits of genetic material, which can be DNA or RNA (ribonucleic acid), surrounded by a protein coat. Viruses are unable to reproduce on their own, and must therefore must infect cells and "hijack" their reproductive apparatus in order to propagate copies of themselves. Antbiotics, as a result, target all manner of bacteria but are ineffective against viruses. Antiviral drugs exist, with newer and more effective ones being introduced all the time, but their mechanisms of action are completely different from those of antibiotics, which usually target either cell walls or metabolic enzymes particular to prokaryotic cells.

The Cell Membrane

The cell membrane is a multifaceted wonder of biology. Its most obvious job is to serve as a container for the contents of the cell and provide a barrier to the insults of the extra-cellular environment. This, however, only describes a small part of its function. The cell membrane is not a passive partition but a highly dynamic assembly of gates and channels that help ensure the maintenance of a cell's internal environment (that is, its equilibrium or homeostasis) by selectively allowing molecules into and out of the cell as required.

The membrane is actually a double membrane, with two layers facing each other in a mirror-image fashion. This is called the phospholipid bilayer, and each layer consists of a "sheet" of phospholipid molecules, or more properly, glycerophospholipid molecules. These are elongated molecules consisting of polar phosphate "heads" that face away from the center of the bilayer (that is, toward the cytoplasm and the cell exterior) and nonpolar "tails" consisting of a pair of fatty acids; these two acids and the phosphate are attached to opposite sides of a three-carbon glycerol molecule. Because of the asymmetrical charge distribution on phosphate groups and the lack of charge asymmetry of fatty acids, phospholipids placed in solution actually assemble themselves spontaneously into this sort of bilayer, so it is energetically efficient.

Substances can traverse the membrane in a variety of ways. One is simple diffusion, which sees small molecules such as oxygen and carbon dioxide move through the membrane from regions of higher concentration to areas of lower concentration. Facilitated diffusion, osmosis and active transport also help maintain a steady supply of nutrients coming into the cell and metabolic waste products exiting.

The Nucleus

The nucleus is the site of DNA storage in eukaryotic cells. (Recall that prokaryotes lack nuclei because they lack membrane-bound organelles of any sort.) Like the plasma membrane, the nuclear membrane, also called a nuclear envelope, is a double-layered phospholipid barrier.

Within the nucleus, the genetic material of a cell is arranged into distinct bodies called chromosomes. The number of chromosomes an organism has varies from species to species; humans have 23 pairs, including 22 pairs of "normal" chromosomes, called autosomes, and one pair of sex chromosomes. The DNA of individual chromosomes is arranged in sequences called genes; each gene carries the genetic code for a particular protein product, be it an enzyme, a contributor to eye color or a component of skeletal muscle.

When a cell undergoes division, its nucleus divides in a distinct way, owing to the replication of the chromosomes within it. This reproductive process is called mitosis, and the cleavage of the nucleus is known as cytokinesis.

Ribosomes

Ribosomes are the site of protein synthesis in cells. These organelles are made almost entirely from a type of RNA fittingly called ribosomal RNA, or rRNA. These ribosomes, which are found throughout the cell cytoplasm, include one large subunit and one small subunit.

Perhaps the easiest way to envision ribosomes is as tiny assembly lines. When it is time to manufacture a given protein product, messenger RNA (mRNA) transcribed in the nucleus from DNA makes its way to the portion of ribosomes where the mRNA code is translated into amino acids, the building blocks of all proteins. Specifically, the four different nitrogenous bases of mRNA can be arranged in 64 different ways into groups of three (4 raised to the third power is 64), and each of these "triplets" codes for an amino acid. Because there are only 20 amino acids in the human body, some amino acids are derived from more than one triplet code.

When the mRNA is being translated, yet another type of RNA, transfer RNA (tRNA) carries whatever amino acid has been summoned by the code to the ribosomal site of synthesis, where the amino acid is attached to the end of the protein-in-progress. Once the protein, which can be anywhere from dozens to many hundreds of amino acids long, is complete, it is released from the ribosome and transported to wherever it is needed.

Mitochondria and Chloroplasts

Mitochondria are the "power plants" of animal cells, and chloroplasts are their analogs in plant cells. Mitochondria, believed to have originated as free-standing bacteria before becoming incorporated into the structures that became eukaryotic cells, are the site of aerobic metabolism, which requires oxygen to extract energy in the form of adenosine triphosphate (ATP) from glucose. The mitochondria receives pyruvate molecules derived from oxygen-independent glucose breakdown in the cytoplasm; in the matrix (interior) of the mitochondria, the pyruvate is subjected to the Krebs cycle, also called the citric-acid cycle or the tricarboxylic acid (TCA) cycle. The Krebs cycle generates a build-up of high-energy proton carriers and serves as set-up for the aerobic reactions called the electron transport chain, which occurs nearby on the mitochondrial membrane, which is yet another lipid bilayer. These reactions generate far more energy in the form of ATP than glycolysis can; without mitochondria, animal life could not have evolved on Earth owing to the prodigious energy requirements of "higher" organisms.

Chloroplasts are what give plants their green color, as they contain a pigment called chlorophyll. Whereas mitochondria breaks down glucose products, chloroplasts actually use the energy from sunlight to build glucose from carbon dioxide and water. The plant then uses some of this fuel for its own needs, but most of it, along with the oxygen liberated in glucose synthesis, reaches the ecosystem and is used by animals, which cannot make their own food. Without abundant plant life on Earth, animals could not survive; the converse is true, as animal metabolism generates sufficient carbon dioxide for plants to use.

The Cytoskeleton

The cytoskeleton, as its name suggests, provides structural support to a cell in the same way your own bony skeleton provides a stable scaffolding for your organs and tissues. The cytoskeleton consists of three components: microfilaments, intermediate fibers and microtubules, in order from smallest to largest. Microfilaments and microtubules can be assembled and disassembled according to the needs of cell at a given time, whereas intermediate filaments tend to be more permanent.

In addition to fixing organelles in place much like the guide wires attached to tall communication towers keep these fixed to the ground, the cytoskeleton assists in moving things within a cell. This can be in the form of serving as anchor points for flagella, as some microtubules do; alternatively, some microtubules provide the actual conduit (pathway) for things to move along. Thus the cytoskeleton can be both motor and highway, depending on the specific type.

Other Organelles

Other important organelles include Golgi bodies, which look like stacks of pancakes on microscopic examination and serve as sites of protein storage and secretion, and the endoplasmic reticulum, which moves protein products along from one portion of the cell to another. Endoplasmic reticulum comes in smooth and rough forms; the latter are so named because they are studded with ribosomes. Golgi bodies give rise to vesicles that break off the edges of the "pancakes" and contain proteins; if these can be regarded as shipping containers, then the endoplasmic reticulum that receives these bodies is like a highway or railroad system.

Lysosomes are also important in the upkeep of cells. These are also vesicles, but they contain specific digestive enzymes that can lyse (dissolve) either the metabolic waste products of cells or chemicals that are not supposed to be there at all but have somehow breached the cell membrane.

Function of Cell

Provide Structure and Support

Like a classroom is made of bricks, every organism is made of cells. While some cells such as the collenchyma and sclerenchyma are specifically meant for structural support, all cells generally provide the structural basis of all organisms. For instance, skin is made up of a number of skin cells. Vascular plants have evolved a special tissue called xylem, which is made of cells that provide structural support.

Facilitate Growth Through Mitosis

In complex organisms, tissues grow by simple multiplication of cells. This takes place through the process of mitosis in which the parent cell breaks down to form two daughter cells identical to it. Mitosis is also the process through which simpler organisms reproduce and give rise to new organisms.

Create Metabolic Reactions

Metabolism includes all the chemical reactions that take place inside an organism to keep it alive. These reactions can be catabolic or anabolic. The process of energy production by breaking down molecules (glucose) is known as catabolism. Anabolic reactions, on the other hand, use energy to make bigger substances from simpler ones.

Energy Generation

Living cells exist in a perpetually active biological state. All cells require energy to conduct the activities necessary to stay alive. This makes energy generation the most essential and fundamental of all cellular functions. Human cells generate energy using chemicals derived from food as fuel. Mitochondria -- often called the powerhouses of cells -- serve as the final site of energy production within human cells. Using molecules derived from the digestion and processing of sugars, proteins and fats from the diet, a series of complex chemical reactions in the mitochondria generate a molecule call adenosine triphosphate (ATP). ATP molecules capture the chemical energy from food, which the cells then use to power their biological functions.

Molecular Transport

Each cell is surrounded by a membrane that delineates its boundaries and acts as a gatekeeper, controlling the movement of molecules into and out of the cell. This molecular transport serves the essential role of maintaining a healthy internal environment so the cell can survive and perform its specialized functions. Some small molecules, such as oxygen, carbon dioxide and water, cross human cell membranes without expending energy. However, the transport of many important molecules depends on energy-utilizing cell membrane pumps. For example, sodium and potassium membrane pumps are essential for the transmission of nerve impulses and maintaining water balance in the human body. Molecular transport also allows specialized cells that produce important chemicals - such as hormones - to release them so they can be utilized elsewhere in the body.

Reproduction

All living things, including cells, have a limited lifespan. Therefore, reproduction is an essential function for cells. Cellular reproduction in humans enables the body to replace dying, diseased or damaged cells -- and in the case of pregnancy, to perpetuate the human species. Most human cells reproduce through a process called mitosis. Using this reproductive process, the cell produces an exact copy of its genetic material before splitting into two genetically identical cells.

Meiosis is a special cellular reproductive process used by eggs and sperm in organisms that employ sexual reproduction. Human genetic material is organized into 23 pairs of chromosomes, one each from the father and the mother. With meiosis, rather than the eggs and sperm having two copies of each chromosome, each contains only a single copy. When the egg and sperm join at the time of fertilization, the chromosomes pair up. Subsequent divisions of the fertilized egg occur via mitosis, with each new cell containing the newly formed pair of chromosomes.

Cell Theory

Cell theory is the idea that the cell is the fundamental structural and functional unit of all living organisms and that new cells are formed from other existing cells. This theory is one of the foundations of modern biology. Today, it is held that all organisms are composed of one or more cells, all vital functions of an organism occur within cells, and cells contain the hereditary information necessary for regulating cell functions and for transmitting information to the next generation of cells.

In a properly functioning multicellular organism, each cell typically has the capacity to provide a useful function for the organism. In turn, the organism provides the necessary nutrients, removal of waste products, and other actions necessary for the survival, growth, and maintenance of the cell during the lifetime of the cell. When cells fail to properly function, such as displaying uncontrolled growth as in cancer, the multicellular organism as a whole can suffer.

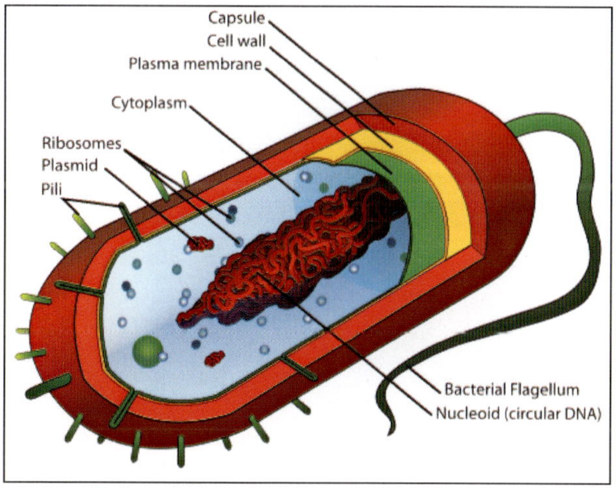

A prokaryote

The principle tenets of cell theory largely were formulated in two publications in the 1800s. In 1838, Matthias Jacob Schleiden, a botanist, published a monograph titled Beiträge zur Phytogenesis in which he proposed that all plant tissues are made from cells. The following year, in 1839, Theodor Schwann, in his publication Mikroskopische Untersuchungen über die Übereinstimmung in der Struktur und dem Wachsthum der Thiere und Pflaznze, generalized this to all living organisms, proposing that all animal tissues also are made from cells and that eggs are cells as well, with all life starting as a single egg. These landmark publications, after subsequent modifications, provided what would become known as the cell theory. According to this theory, the cell is the minimum quantity of life, capable of independent development, yet an integral part of multicellular organisms; "an individual in a society" . Prior to the development of this theory, people recognized tissues and organs, but considered them to be simple substances similar to nonliving materials, not recognizing the level of organization between simple substances and tissues.

The foundation for the cell theory began two centuries earlier, in the 1600s. In 1665, Robert Hooke published his work Micrographia, which contained a number of microscopic observations, and in which he coined the term cell, so called because his observations of plant cells reminded him of the small rooms that monks inhabited, which were called "cellula." Hooke is often credited with the discovery of the cell. However what Hooke actually saw was the non-living cell walls from cork. The cell walls observed by Hooke gave no indication of the nucleus and other organelles found in most living cells. In his observations, Hook did not indicate that he was making a fundamental discovery about life.

The first man to witness a live cell under a microscope was Antonie van Leeuwenhoek, who in 1674 described the algae Spirogyra and named the moving organisms animalcules, meaning "little animals." Leeuwenhoek probably also saw bacteria.

The idea that cells were separable into individual units was proposed by Ludolph Christian Treviranus and Johann Jacob Paul Moldenhawer. All of this finally led to Henri Dutrochet formulating one of the fundamental tenets of modern cell theory by declaring that "the cell is the fundamental element of organization" . In 1831, Robert Brown first noticed that plant cells all contain a nucleus, and in 1835 Johannes Purkinje noted that animal tissues also are made of cells, although these observations did not attract attention to the importance of cells.

Recognition of the importance of cells changed quickly in 1835 with Schleiden's proposal that all plant tissues are made of cells and Schwann recognized that all animals tissues are made from cells and suggested all life starts as a single cell. Schleiden and Schwann generally are credited with the cell theory because both contributed vital parts of the theory. Their theory accepted the first two tenets of classical cell theory, that living things are composed of cells and that cells are the basic unit of structure and function in living things. However, the cell theory of Schleiden differed from modern cell theory in that it proposed a method of spontaneous crystallization that he called "free cell formation," rather than the view that all cells are produced from other cells.

Subsequent discoveries extended the cell theory. Karl Wilhelm von Nägeli showed that cells form by division of one cell into two, and in 1845 Karl Theodor Ernst von Siebold extended the cell theory to protists. In the 1840s, Rudolf Albert von Köllikeer reported that sperm are cells and that nerve fibers are part of cells.

In the 1850s, Rudolph Carl Virchow promoted the cell theory and had the credo "all cells arise from cells". Virchow's contribution in 1858 that all cells come from pre-existing cells completed the third component of classical cell theory. Thus, Virchow often is credited with cell theory along with Schleiden and Schwann.

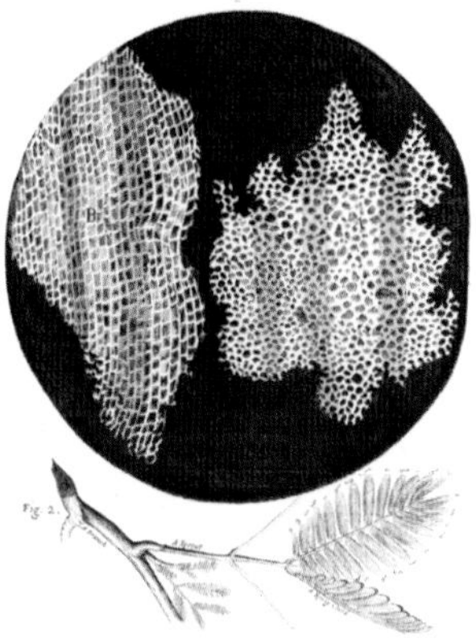

Drawing of the structure of cork

Tenets of Cell Theory

Cell theory is a widely accepted explanation of the relationship between cells and living things. Cell theory holds true for all living things, no matter how big or small, or how simple or complex. Since according to research, cells are common to all living things, they can provide information about all life. And because all cells come from other cells, scientists can study cells to learn about growth, reproduction, and all other functions that living things perform. By learning about cells and how they function, you can learn about all types of living things.

Classical Cell Theory

Classical cell theory, as developed through the observations of Hooke, Leeuwenhoek, Schleiden, Schwann, Virchow, and others, holds that:

1. All organisms are made up of one or more cells.
2. Cells are the fundamental functional and structural unit of life.
3. All cells come from pre-existing cells.

Modern Cell Theory

The generally accepted parts of modern cell theory include:

1. The cell is the fundamental unit of structure and function in living things.
2. All cells come from pre-existing cells by division.
3. Energy flow (metabolism and biochemistry) occurs within cells.
4. Cells contain hereditary information (DNA) which is passed from cell to cell during cell division.

5. All cells are basically the same in chemical composition.
6. All known living things are made up of cells.
7. Some organisms are unicellular, made up of only one cell.
8. Other organisms are multicellular, composed of countless number of cells.
9. The activity of an organism depends on the total activity of independent cells.

Exceptions to the Theory

1. Viruses are considered by some to be alive, yet they are not made up of cells.
2. The first cell did not originate from a pre-existing cell.

Types of Cells

Cells can be subdivided into two basic categories: prokaryotes and eukaryotes. Prokaryotes lack a nucleus (though they do have circular DNA) and most other membrane-bound organelles (though they do contain ribosomes). Bacteria and Archaea are two divisions of prokaryotes. Eukaryotes, on the other hand, have distinct nuclei and membrane-bound organelles (mitochondria, chloroplasts, lysosomes, rough and smooth endoplasmic reticulum, vacuoles). In addition, they possess organized chromosomes that store genetic material.

Prokaryotic Cells

Prokaryotes are single-celled organisms that are the earliest and most primitive forms of life on earth. As organized in the Three Domain System, prokaryotes include bacteria and archaeans. Some prokaryotes, such as cyanobacteria, are photosynthetic organisms and are capable of photosynthesis.

Many prokaryotes are extremophiles and are able to live and thrive in various types of extreme environments including hydrothermal vents, hot springs, swamps, wetlands, and the guts of humans and animals (Helicobacter pylori). Prokaryotic bacteria can be found almost anywhere and are part of the human microbiota. They live on your skin, in your body, and on everyday objects in your environment.

Components of Prokaryotic Cells

There are some key ingredients that a cell needs in order to be a cell, regardless of whether it is prokaryotic or eukaryotic. All cells share four key components:

1. The plasma membrane is an outer covering that separates the cell's interior from its surrounding environment.
2. Cytoplasm consists of the jelly-like cytosol inside the cell, plus the cellular structures suspended in it. In eukaryotes, cytoplasm specifically means the region outside the nucleus but inside the plasma membrane.

3. DNA is the genetic material of the cell.
4. Ribosomes are molecular machines that synthesize proteins.

Despite these similarities, prokaryotes and eukaryotes differ in a number of important ways. A prokaryote is a simple, single-celled organism that lacks a nucleus and membrane-bound organelles.

The majority of prokaryotic DNA is found in a central region of the cell called the nucleoid, and it typically consists of a single large loop called a circular chromosome. The nucleoid and some other frequently seen features of prokaryotes are shown in the diagram below of a cut-away of a rod-shaped bacterium.

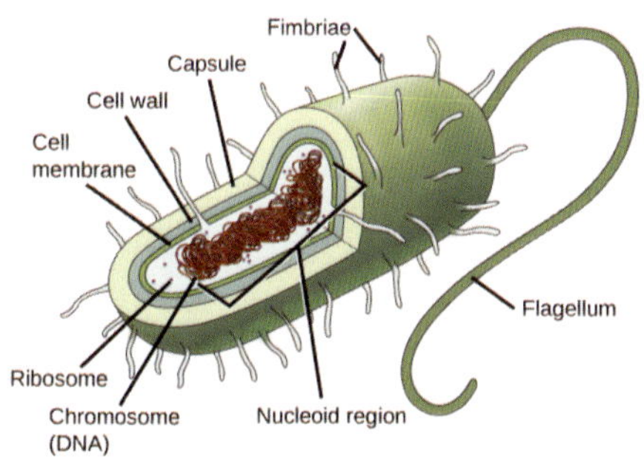

Figure: Prokaryotic cells

Bacteria are very diverse in form, so not every type of bacterium will have all of the features shown in the diagram.

Most bacteria are, however, surrounded by a rigid cell wall made out of peptidoglycan, a polymer composed of linked carbohydrates and small proteins. The cell wall provides an extra layer of protection, helps the cell maintain its shape, and prevents dehydration. Many bacteria also have an outermost layer of carbohydrates called the capsule. The capsule is sticky and helps the cell attach to surfaces in its environment.

Some bacteria also have specialized structures found on the cell surface, which may help them move, stick to surfaces, or even exchange genetic material with other bacteria. For instance, flagella are whip-like structures that act as rotary motors to help bacteria move.

Fimbriae are numerous, hair-like structures that are used for attachment to host cells and other surfaces. Bacteria may also have rod-like structures known as pili, which come in different varieties. For instance, some types of pili allow a bacterium to transfer DNA molecules to other bacteria, while others are involved in bacterial locomotion—helping the bacterium move.

Archaea may also have most of these cell surface features, but their versions of a particular feature are typically different from those of bacteria. For instance, although archaea also have a cell wall, it's not made out of peptidoglycan—although it does contain carbohydrates and proteins.

Size of Prokaryotics Cells

Prokaryotic cells are tiny. They are about 10% as wide as the eukaryotic cells of plants, animals or fungi. The diameter of a prokaryotic cell is usually between 1-10 μm, whereas a typical eukaryotic cell is between 10-100 μm.

The smallest bacteria, called mycoplasmas, can be as little as 0.1 μm in diameter. When you consider that 1000 μm = 1 mm or 0.4 in, then it would take 25,000 mycoplasmas lined up next to each other to make a line 1 inch long (2.5 cm long).

Prokaryotic cells have never evolved to produce multicellular organisms as eukaryotic cells have. All prokaryotes are single-celled microorganisms. Often cells aggregate together to form colonies. A colony of prokaryotic cells often make changes to their surrounding environments that benefits the individual cells.

The size of a prokaryotic cell is restricted by its metabolism. Because prokaryotic cells lack organelles, their cellular processes aren't as efficient as eukaryotic cells, which is why they are unable to grow as large.

Increases in the size of a cell lead to increases in nutrient and energy demands. Nutrients enter a cell through the plasma membrane but as a cell gets bigger, the volume of the cell increases faster than the surface area of the plasma membrane and energy demands increase more than the energy supply.

To counteract the difference between the supply and demand of nutrients and energy, cells must become more efficient at using their supply of nutrients and energy. Eukaryotic cells with organelles have evolved far more efficient methods for using their supply of nutrients and energy and so they have been able to grow much larger than prokaryotic cells.

Metabolism of Prokaryotic Cells

Prokaryotic organisms have evolved a wide range of ways of to take energy from the environment. Compared to eukaryotic cells that have only evolved to transfer energy through photosynthesis and respiration, prokaryotic cells can obtain energy through photosynthesis, respiration, nitrogen fixation, denitrification, sulfate reduction and methanogenesis. These words may not mean anything to you but they illustrate the diversity of ways prokaryotic cells can take energy from their surrounding environment.

Importance of Prokaryotic cell

Prokaryotes are important to all life on Earth for a number of reasons. They play a critical role in the recycling of nutrients by decomposing dead organisms and allowing their nutrients to be re-used.

They are also important for many metabolic processes. Bacteria in our guts and mouth help with the digestion of food by breaking down difficult to digest carbohydrates and other compounds. Recent research has suggested that the microbial community in our guts can be a significant factor contributing to obesity.

Binary Fission

Most prokaryotes reproduce asexually through a process called binary fission. During binary fission, the single DNA molecule replicates and the original cell is divided into two identical cells.

Steps of binary fission:

- Binary fission begins with DNA replication of the single DNA molecule. Both copies of DNA attach to the cell membrane.
- Next, the cell membrane begins to grow between the two DNA molecules. Once the bacterium just about doubles its original size, the cell membrane begins to pinch inward.
- A cell wall then forms between the two DNA molecules dividing the original cell into two identical daughter cells.

Although E.coli and other bacteria most commonly reproduce by binary fission, this mode of reproduction does not produce genetic variation within the organism.

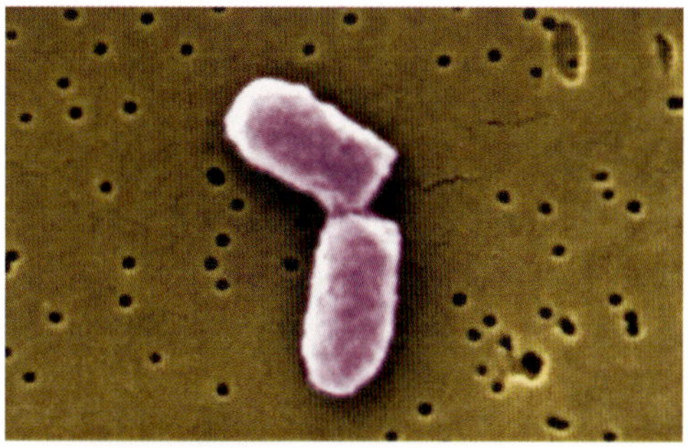

E. coli bacteria undergoing binary fission. The cell wall is dividing resulting in the formation of two cells.

Prokaryotic Recombination

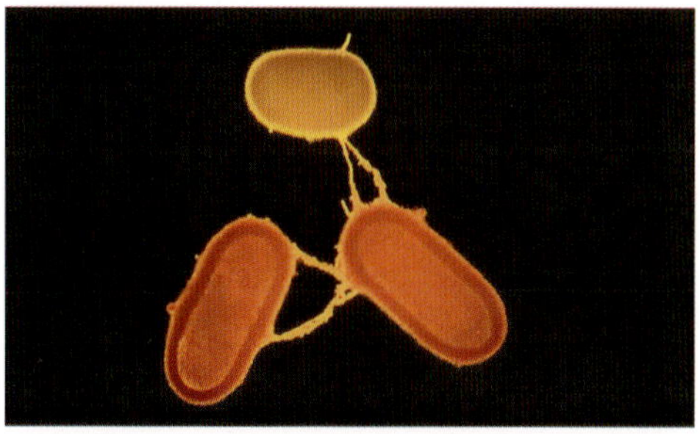

False-color transmission electron micrograph (TEM) of an Escherichia coli bacterium (bottom right) conjugating with two other E.coli bacteria. The tubes connecting the bacteria are pili, which are used to transfer genetic material between bacteria.

Genetic variation within prokaryotic organisms is accomplished through recombination. In recombination, genes from one prokaryote are incorporated into the genome of another prokaryote. Recombination is accomplished in bacterial reproduction by the processes of conjugation, transformation, or transduction.

- In conjugation, bacteria connect with one another through a protein tube structure called a pilus. Genes are transferred between bacteria through the pilus.
- In transformation, bacteria take up DNA from their surrounding environment. The DNA is transported across the bacterial cell membrane and incorporated into the bacterial cell's DNA.

- Transduction involves the exchange of bacterial DNA through viral infection. Bacteriophages, viruses that infect bacteria, transfer bacterial DNA from previously infected bacteria to any additional bacteria that they infect.

Eukaryotic Cells

Eukaryotic cells are cells that contain a nucleus and organelles, and are enclosed by a plasma membrane. Organisms that have eukaryotic cells include protozoa, fungi, plants and animals. These organisms are grouped into the biological domain Eukaryota. Eukaryotic cells are larger and more complex than prokaryotic cells, which are found in Archaea and Bacteria, the other two domains of life.

Structure of Eukaryotic Cells

The Plasma Membrane

Like bacteria and archaea, eukaryotic cells have a plasma membrane, a phospholipid bilayer with embedded proteins that separates the internal contents of the cell from its surrounding environment. The plasma membrane controls the passage of organic molecules, ions, water, and oxygen into and out of the cell. Wastes (such as carbon dioxide and ammonia) also leave the cell by passing through the plasma membrane, usually with some help of protein transporters.

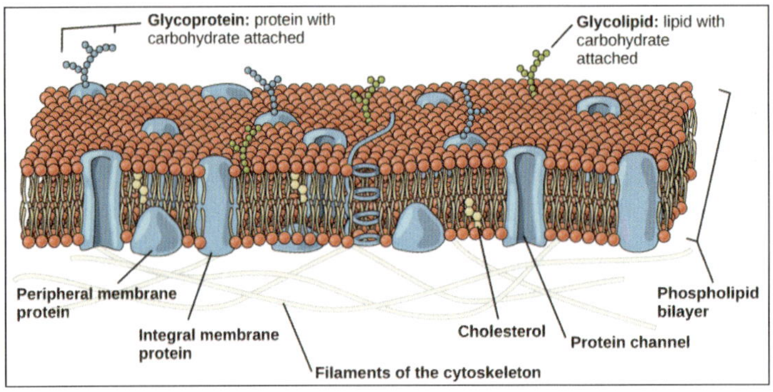

Figure: The eukaryotic plasma membrane is a phospholipid bilayer with proteins and cholesterol embedded in it..

In the context of bacterial cell membranes, the plasma membranes of eukaryotic cells may also adopt unique conformations. For instance, the plasma membrane of cells that, in multicellular organisms, specialize in absorption are often folded into fingerlike projections called microvilli (singular = microvillus). The "folding" of the membrane into microvilli effectively increases the surface area for absorption while minimally impacting the cytosolic volume. Such cells can be found lining the small intestine, the organ that absorbs nutrients from digested food.

People with celiac disease have an immune response to gluten, a protein found in wheat, barley, and rye. The immune response damages microvilli. As a consequence, afflicted individuals have an impaired ability to absorb nutrients. This can lead to malnutrition, cramping, and diarrhea.

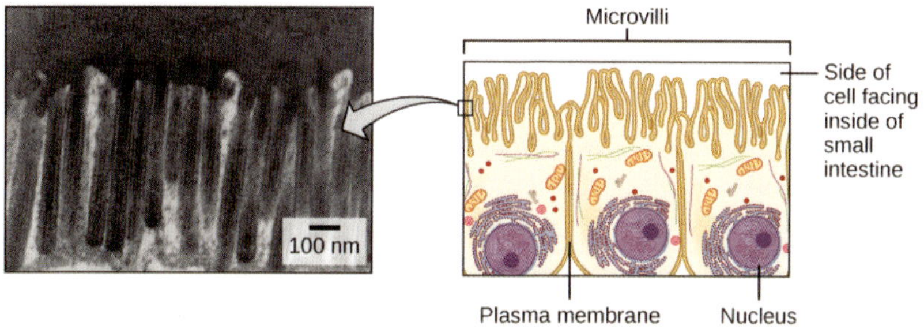

Microvilli shown in figure appear on cells lining the small intestine, increase the surface area available for absorption. These microvilli are only found on the area of the plasma membrane that faces the cavity from which substances will be absorbed.

The Cytoplasm

The cytoplasm refers to the entire region of a cell between the plasma membrane and the nuclear envelope. It is composed of organelles suspended in the gel-like cytosol, the cytoskeleton, and various chemicals (figure). Even though the cytoplasm consists of 70 to 80 percent water, it nevertheless has a semisolid consistency. It is crowded in there. Proteins, simple sugars, polysaccharides, amino acids, nucleic acids, fatty acids, ions and many other water-soluble molecules are all competing for space and water.

The Nucleus

Typically, the nucleus is the most prominent organelle in a cell (figure below) when viewed through a microscope. The nucleus (plural = nuclei) houses the cell's DNA.

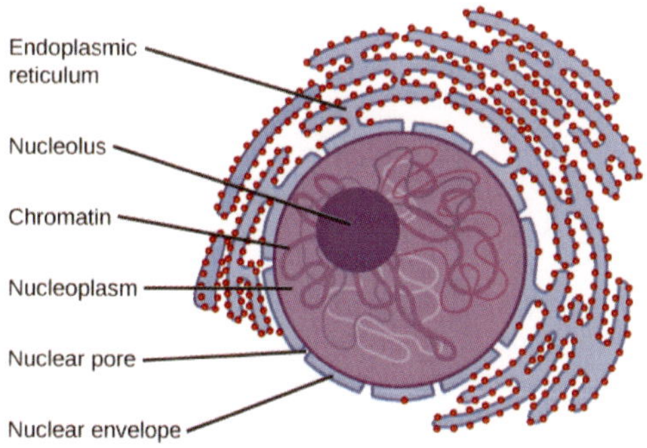

The nucleus stores chromatin (DNA plus proteins) in a gel-like substance called the nucleoplasm. The nucleolus is a condensed region of chromatin where ribosome synthesis occurs. The boundary of the nucleus is called the nuclear envelope. It consists of two phospholipid bilayers: an outer membrane and an inner membrane. The nuclear membrane is continuous with the endoplasmic reticulum. Nuclear pores allow substances to enter and exit the nucleus.

The Nuclear Envelope

The nuclear envelope, a structure that constitutes the outermost boundary of the nucleus, is a double-membrane—both the inner and outer membranes of the nuclear envelope are phospholipid bilayers. The nuclear envelope is also punctuated with protein-based pores that control the passage of ions, molecules, and RNA between the nucleoplasm and cytoplasm. The nucleoplasm is the semisolid fluid inside the nucleus where we find the chromatin and the nucleolus, a condensed region of chromatin where ribosome synthesis occurs.

Chromatin and Chromosomes

To understand chromatin, it is helpful to first consider chromosomes. Chromosomes are structures within the nucleus that are made up of DNA, the hereditary material. You may remember that in bacteria and archaea, DNA is typically organized into one or more circular chromosome(s). In eukaryotes, chromosomes are linear structures. Every eukaryotic species has a specific number of chromosomes in the nuclei of its cells. In humans, for example, the chromosome number is 23, while in fruit flies, it is 4.

Chromosomes are only clearly visible and distinguishable from one another by visible optical microscopy when the cell is preparing to divide and the DNA is tightly packed by proteins into easily distinguishable shapes. When the cell is in the growth and maintenance phases of its life cycle, numerous proteins are still associated with the nucleic acids, but the DNA strands more closely resemble an unwound, jumbled bunch of threads. The term chromatin is used to describe chromosomes (the protein-DNA complexes) when they are both condensed and decondensed.

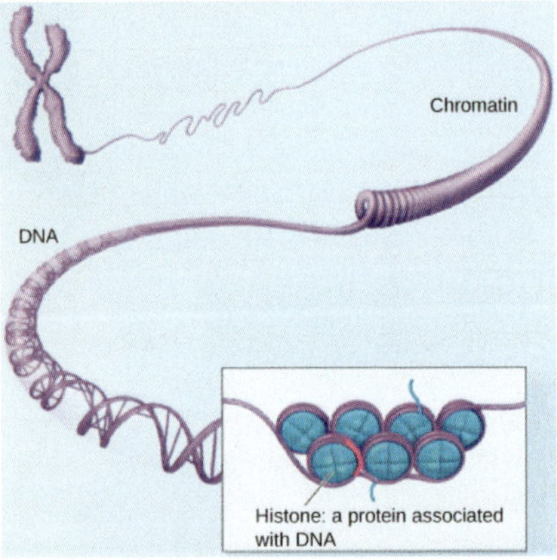

Figure: (a) This image shows various levels of the organization of chromatin (DNA and protein). (b) This image shows paired chromosomes.

The Nucleolus

Some chromosomes have sections of DNA that encode ribosomal RNA. A darkly staining area within the nucleus called the nucleolus (plural = nucleoli) aggregates the ribosomal RNA with associated proteins to assemble the ribosomal subunits that are then transported out to the cytoplasm through the pores in the nuclear envelope.

Ribosomes

Ribosomes are the cellular structures responsible for protein synthesis. When viewed through an electron microscope, ribosomes appear either as clusters (polyribosomes) or single, tiny dots that float freely in the cytoplasm. They may be attached to the cytoplasmic side of the plasma membrane or the cytoplasmic side of the endoplasmic reticulum and the outer membrane of the nuclear envelope.

Electron microscopy has shown us that ribosomes, which are large complexes of protein and RNA, consist of two subunits, aptly called large and small (figure below). Ribosomes receive their "instructions" for protein synthesis from the nucleus, where the DNA is transcribed into messenger RNA (mRNA). The mRNA travels to the ribosomes, which translate the code provided by the sequence of the nitrogenous bases in the mRNA into a specific order of amino acids in a protein. This is covered in greater detail in the section covering the process of translation.

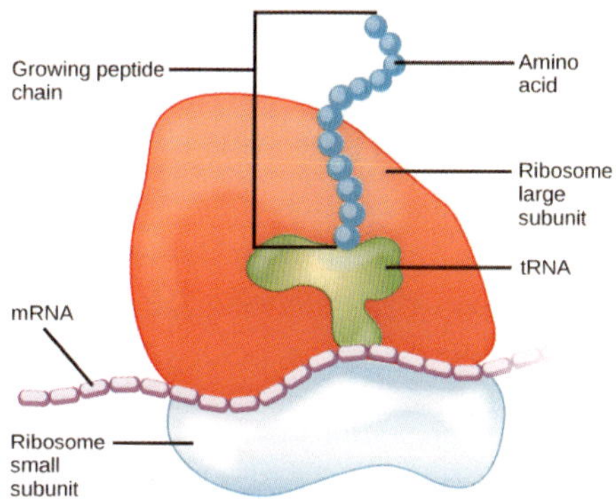

Figure: Ribosomes are made up of a large subunit (top) and a small subunit (bottom). During protein synthesis, ribosomes assemble amino acids into proteins.

Mitochondria

Mitochondria (singular = mitochondrion) are often called the "powerhouses" or "energy factories" of a cell because they are the primary site of metabolic respiration in eukaryotes. Depending on the species and the type of mitochondria found in those cells, the respiratory pathways may be anaerobic or aerobic. By definition, when respiration is aerobic, the terminal electron is oxygen; when respiration is anaerobic, a compound other than oxygen functions as the terminal electron acceptor. In either case, the result of these respiratory processes is the production of ATP via oxidative phosphorylation, hence the use of terms "powerhouse" and/or "energy factory" to describe this organelle. Nearly all mitochondria also possess a small genome that encodes genes whose functions are typically restricted to the mitochondrion.

In some cases, the number of mitochondria per cell is tunable, depending, typically, on energy demand. It is for instance possible muscle cells that are used—that by extension have a higher demand for ATP—may often be found to have a significantly higher number of mitochondria than cells that do not have a high energy load.

The structure of the mitochondria can vary significantly depending on the organism and the state of the cell cycle which one is observing. However, depicts mitochondria as oval-shaped organelles with a double inner and outer membrane (figure below); learn to recognize this generic representation. Both the inner and outer membranes are phospholipid bilayers embedded with proteins that mediate transport across them and catalyze various other biochemical reactions. The inner membrane layer has folds called cristae that increase the surface area into which respiratory chain proteins can be embedded. The region within the cristae is called the mitochondrial matrix and contains—among other things—enzymes of the TCA cycle. During respiration, protons are pumped by respiratory chain complexes from the matrix into a region known as the inter membrane space (between the inner and outer membranes).

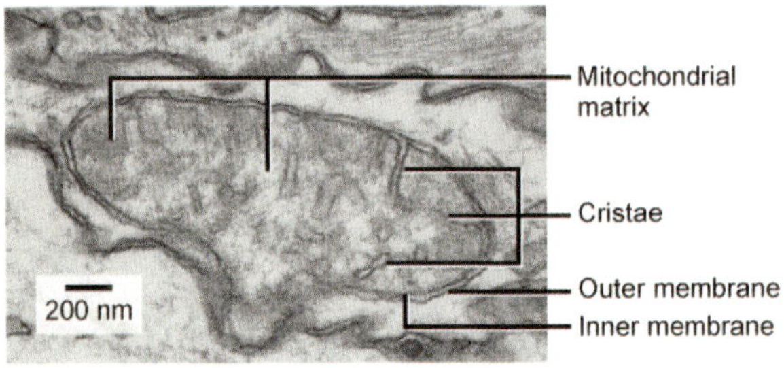

This electron micrograph shows a mitochondrion as viewed with a transmission electron microscope. This organelle has an outer membrane and an inner membrane. The inner membrane contains folds, called cristae, which increase its surface area. The space between the two membranes is called the intermembrane space, and the space inside the inner membrane is called the mitochondrial matrix. ATP synthesis takes place on the inner membrane.

Peroxisomes

Peroxisomes are small, round organelles enclosed by single membranes. These organelles carry out redox reactions that oxidize and break down fatty acids and amino acids. They also help to detoxify many toxins that may enter the body. Many of these redox reactions release hydrogen peroxide, H_2O_2, which would be damaging to cells; however, when these reactions are confined to peroxisomes, enzymes safely break down the H_2O_2 into oxygen and water. For example, alcohol is detoxified by peroxisomes in liver cells. Glyoxysomes, which are specialized peroxisomes in plants, are responsible for converting stored fats into sugars.

Vesicles and Vacuoles

Vesicles and vacuoles are membrane-bound sacs that function in storage and transport. Other than the fact that vacuoles are somewhat larger than vesicles, there is a very subtle distinction between them: the membranes of vesicles can fuse with either the plasma membrane or other membrane systems within the cell. Additionally, some agents such as enzymes within plant vacuoles break down macromolecules. The membrane of a vacuole does not fuse with the membranes of other cellular components.

Animal Cells Versus Plant Cells

At this point, you know that each eukaryotic cell has a plasma membrane, cytoplasm, a nucleus, ribosomes, mitochondria, peroxisomes, and in some, vacuoles. There are some striking differences between animal and plant cells worth noting. Here is a brief list of differences that we want you to be familiar with and a slightly expanded description below:

1. While all eukaryotic cells use microtubule and motor protein the based mechanisms to segregate chromosomes during cell division, the structures used to organize these microtubules differ in plants versus animal and yeast cells. Animal and yeast cells organize and anchor their microtubules into structures called microtubule organizing centers (MTOCs).

These structures are composed of structures called centrioles that are composed largely of α-tubulin, β-tubulin, and other proteins. Two centrioles organize into a structure called a centrosome. By contrast, in plants, while microtubules also organize into discrete bundles, there are no conspicuous structures similar to the MTOCs seen in animal and yeast cells. Rather, depending on the organism, it appears that there can be several places where these bundles of microtubules can nucleate from places called acentriolar (without centriole) microtubule organizing centers. A third type of tubulin, γ-tubulin, appears to be implicated, but our knowledge of the precise mechanisms used by plants to organize microtubule spindles is still spotty.

2. Animal cells typically have organelles called lysosomes responsible for degradation of biomolecules. Some plant cells contain functionally similar degradative organelles, but there is a debate as to how they should be named. Some plant biologists call these organelles lysosomes while others lump them into the general category of plastids and do not give them a specific name.
3. Plant cells have a cell wall, chloroplasts and other specialized plastids, and a large central vacuole, whereas animal cells do not.

The Centrosome

The centrosome is a microtubule-organizing center found near the nuclei of animal cells. It contains a pair of centrioles, two structures that lie perpendicular to eachother (figure below). Each centriole is a cylinder of nine triplets of microtubules.

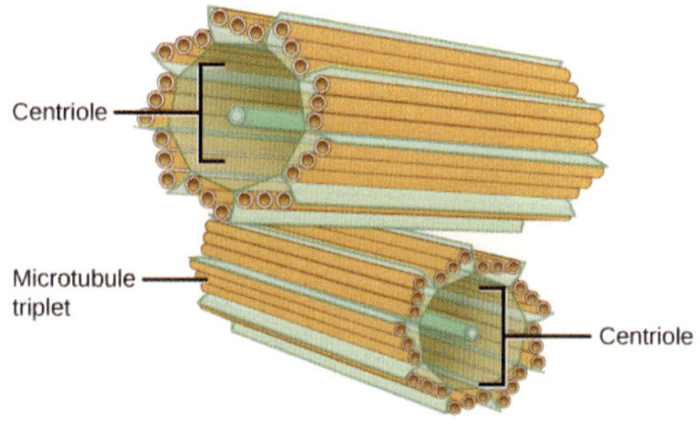

Figure: The centrosome consists of two centrioles that lie at right angles to each other. Each centriole is a cylinder made up of nine triplets of microtubules. Nontubulin proteins (indicated by the green lines) hold the microtubule triplets together.

The centrosome (the organelle where all microtubules originate in animal and yeast) replicates itself before a cell divides, and the centrioles appear to have some role in pulling the duplicated chromosomes to opposite ends of the dividing cell. However, the exact function of the centrioles in cell division remains unclear, as cells that have had their centrosome removed can still divide, and plant cells, which lack centrosomes, are capable of cell division.

Lysosomes

Animal cells have another set of organelles not found in plant cells: lysosomes. Colloquially, the lysosomes are sometimes called the cell's "garbage disposal". Enzymes within the lysosomes aid the breakdown of proteins, polysaccharides, lipids, nucleic acids, and even "worn-out" organelles.

These enzymes are active at a much lower pH than that of the cytoplasm. Therefore, the pH within lysosomes is more acidic than the pH of the cytoplasm. In plant cells, many of the same digestive processes take place in vacuoles.

The Cell Wall

If you examine the diagram above depicting plant and animal cells, you will see in the diagram of a plant cell a structure external to the plasma membrane called the cell wall. The cell wall is a rigid covering that protects the cell, provides structural support, and gives shape to the cell. Fungal and protistan cells also have cell walls. While the chief component of bacterial cell walls is peptidoglycan, the major organic molecule in the plant cell wall is cellulose a polysaccharide made up of glucose subunits.

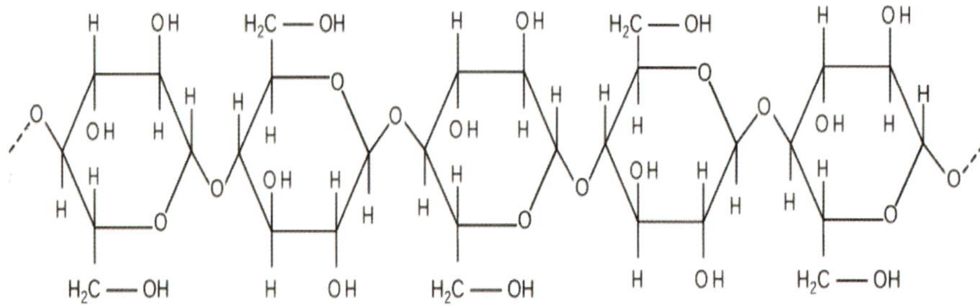

Figure: Cellulose is a long chain of β-glucose molecules connected by a 1-4 linkage. The dashed lines at each end of the figure indicate a series of many more glucose units. The size of the page makes it impossible to portray an entire cellulose molecule.

Chloroplasts

Chloroplasts are plant cell organelles that carry out photosynthesis. Like the mitochondria, chloroplasts have their own DNA and ribosomes, but chloroplasts have an entirely different function.

Like mitochondria, chloroplasts have outer and inner membranes, but within the space enclosed by a chloroplast's inner membrane is a set of interconnected and stacked fluid-filled membrane sacs called thylakoids (figure below). Each stack of thylakoids is called a granum (plural = grana). The fluid enclosed by the inner membrane that surrounds the grana is called the stroma.

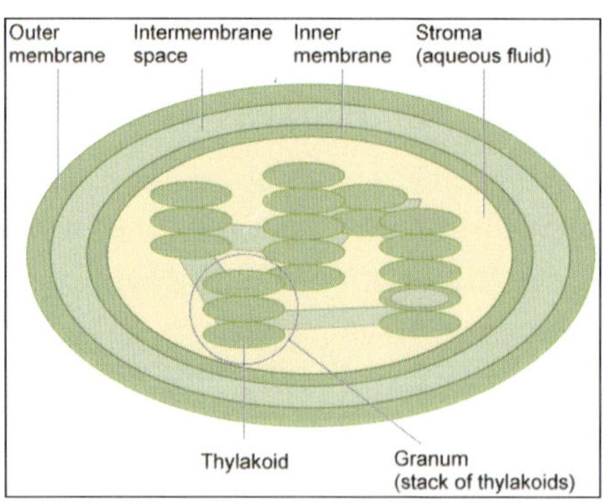

The chloroplast has an outer membrane, an inner membrane, and membrane structures called thylakoids that are stacked into grana. The space inside the thylakoid membranes is called the thylakoid space. The light harvesting reactions take place in the thylakoid membranes, and the synthesis of sugar takes place in the fluid inside the inner membrane, which is called the stroma. Chloroplasts also have their own genome, which is contained on a single circular chromosome.

The chloroplasts contain a green pigment called chlorophyll, which captures the light energy that drives the reactions of photosynthesis. Like plant cells, photosynthetic protists also have chloroplasts. Some bacteria perform photosynthesis, but their chlorophyll is not relegated to an organelle.

The Central Vacuole

The central vacuole plays a key role in regulating the cell's concentration of water in changing environmental conditions.

Silly vacuole factoid: Have you ever noticed that if you forget to water a plant for a few days, it wilts? That's because as the water concentration in the soil becomes lower than the water concentration in the plant, water moves out of the central vacuoles and cytoplasm. As the central vacuole shrinks, it leaves the cell wall unsupported. This loss of support to the cell walls of plant cells results in the wilted appearance of the plant.

The central vacuole also supports the expansion of the cell. When the central vacuole holds more water, the cell gets larger without having to invest a lot of energy in synthesizing new cytoplasm.

Characteristics of Eukaryotic Cells

Eukaryotic cells contain a variety of structures called organelles, which perform various functions within the cell. Examples of organelles are ribosomes, which make proteins, the endoplasmic reticulum, which sorts and packages the proteins, and mitochondria, which produce the energy molecule adenosine triphosphate (ATP). They also have a true nucleus, which contains the genetic material DNA and is surrounded by a nuclear envelope. All of the organelles are stabilized and given physical support through the cytoskeleton, which is also involved in sending signals from one part of the cell to the other. In eukaryotic cells, the cytoskeleton is composed mainly of three types of filaments: microtubules, microfilaments, and intermediate filaments. The gel-like substance that surrounds all the organelles in the cell is called cytosol.

The figure below shows the structure of a eukaryotic cell; this is an animal cell. The nucleus and other organelles are shown. The cytosol is the blue substance surrounding all of the organelles. Together, the cytosol with all organelles besides the nucleus are grouped as cytoplasm.

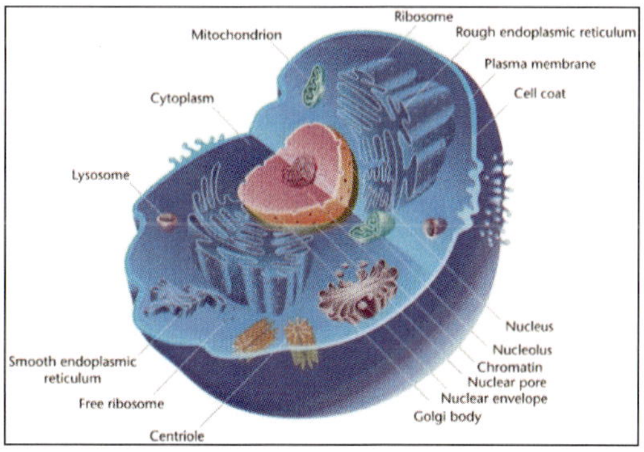

Eukaryotic Cell Cycle

The cell cycle is the life cycle of a cell. During this cycle, it grows and divides. Checkpoints exist between all stages so that proteins can determine whether the cell is ready to begin the next phase of the cycle.

Quiescence (G_0)

Quiescence, also known as senescence or resting, is a phase in which the cell is not actively dividing. It is also known as Gap 0, or G_0. This stage is considered the start of the cell cycle, although it is one that cells can reach and then stop dividing indefinitely, which ends the cell cycle. Liver, stomach, kidney cells, and neuron are all examples of cells that can reach this stage and remain in it for long periods of time. It can also occur when a cell's DNA is damaged. However, most cells do not go into the G_0 stage at all, and can divide indefinitely throughout the life of an organism.

Interphase

In interphase, the cell grows and takes in nutrients in preparation for division. Interphase takes up about 90 percent of the cell cycle. It consists of three parts: Gap 1, Synthesis, and Gap 2.

- Gap 1 (G1) is also known as a growth phase. The cell gets larger and increases its stock of proteins, along with organelles such as the energy-producing mitochondria.
- Synthesis (S) is the phase in which DNA replicates. During synthesis, the chromosomes replicate so that each chromosome is made up of two sister chromatids. At the end of this phase, there is double the amount of DNA in the cell.
- Gap 2 (G_2) is another growth phase. The cell becomes even larger in order to prepare for mitotic division.

Mitosis (M)

Mitosis, or M phase, is when the cell begins to organize its duplicated DNA for separation into two daughter cells. The chromosomes separate so that one of each chromosome goes into each daughter cell. This results in the daughter cells having identical chromosomes to the parent cell. Mitosis itself is divided into prophase, metaphase, anaphase, and telophase, which mark various points in the DNA separation process. Mitosis is then followed by a process called cytokinesis, during which the cell separates its nuclei and other organelles in preparation for division and then physically divides into two cells.

Examples of Eukaryotic Cells

Plant Cells

Plant cells are unique among eukaryotic cells for several reasons. They have reinforced, relatively thick cell walls that are made mostly of cellulose and help maintain structural support in the plant. Each plant cell has a large vacuole in the center that allows it to maintain turgor pressure, which is pressure from having a lot of water in the cell and helps keep the plant upright. Plant cells also contain organelles called chloroplasts which contain the molecule chlorophyll. This important molecule is used in the process of photosynthesis, which is when a plant makes its own energy from sunlight, carbon dioxide, and water.

Fungal Cells

Like plant cells, fungal cells also have a cell wall, but their cell wall is made of chitin (the same substance found in insect exoskeletons). Some fungi have septa, which are holes that allow organelles and cytoplasm to pass between them. This makes the boundaries between different cells less clear.

Animal Cells

Animal cells do not have cell walls. Instead, they have only a plasma membrane. The lack of a cell wall allows animal cells to form many different shapes, and allows for the processes of phagocytosis "cell eating" and pinocytosis "cell drinking" to occur. Animal cells differ from plant cells in that they do not have chloroplasts and have smaller vacuoles instead of a large central vacuole.

Protozoa

Protozoa are eukaryotic organisms that consist of a single cell. They can move around and eat, and they digest food in vacuoles. Some protozoa have many cilia, which are small "arms" that allow them to move around. Some also have a thin layer called a pellicle, which provides support to the cell membrane.

Cellular Process

Cell Division

Cell division is the process cells go through to divide. There are several types of cell division, depending upon what type of organism is dividing. Organisms have evolved over time to have different and more complex forms of cell division. Most prokaryotes, or bacteria, use binary fission to divide the cell. Eukaryotes of all sizes use mitosis to divide. Sexually-reproducing eukaryotes use a special form of cell division called meiosis to reduce the genetic content in the cell. This is necessary in sexual reproduction because each parent must give only half of the required genetic material, otherwise the offspring would have too much DNA, which can be a problem.

Types of Cell Division

Prokaryotic Cell Division

Prokaryotes replicate through a type of cell division known as binary fission. Prokaryotes are simple organism, with only one membrane and no division internally. Thus, when a prokaryote divides, it simply replicates the DNA and splits in half. The process is a little more complicated than this, as DNA must first be unwound by special proteins. Although the DNA in prokaryotes usually exists in a ring, it can get quite tangled when it is being used by the cell. To copy the DNA efficiently, it must be stretched out. This also allows the two new rings of DNA created to be separated after they are produced. The two strands of DNA separate into two different sides of the prokaryote cell. The cell then gets longer, and divides in the middle. The process can be seen in the figure below.

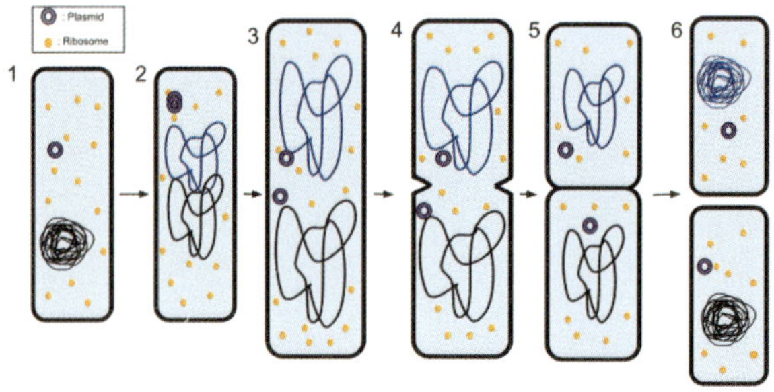

The DNA is the tangled line. The other components are labeled. Plasmids are small rings of DNA that also get copied during binary fission and can be picked up in the environment, from dead cells that break apart. These plasmids can then be further replicated. If a plasmid is beneficial, it will increase in a population. This is in part how antibiotic resistance in bacteria happens. The ribosomes are small protein structures that help produce proteins. They are also replicated so each cell can have enough to function.

Eukaryotic Cell Division: Mitosis

Eukaryotic organisms have membrane bound organelles and DNA that exists on chromosomes, which makes cell division harder. Eukaryotes must replicate their DNA, organelles, and cell mechanisms before dividing. Many of the organelles divide using a process that is essentially binary fission, leading scientist to believe that eukaryotes were formed by prokaryotes living inside of other prokaryotes.

After the DNA and organelles are replicated during interphase of the cell cycle, the eukaryote can begin the process of mitosis. The process begins during prophase, when the chromosomes condense. If mitosis proceeded without the chromosomes condensing, the DNA would become tangled and break. Eukaryotic DNA is associated with many proteins which can fold it into complex structures. As mitosis proceeds to metaphase the chromosomes are lined up in the middle of the cell. Each half of a chromosome, known as sister chromatids because they are replicated copies of each other, gets separated into each half of the cell as mitosis proceeds. At the end of mitosis, another process called cytokinesis divides the cell into two new daughter cells.

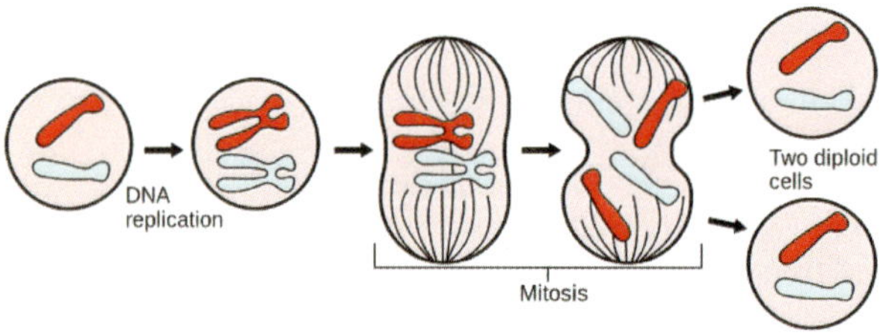

All eukaryotic organisms use mitosis to divide their cells. However, only single-celled organisms use mitosis as a form of reproduction. Most multicellular organisms are sexually reproducing and combine their DNA with that of another organism to reproduce. In these cases, organisms need a different method of cell division. Mitosis yields identical cells, but meiosis produces cells with half the genetic information of a regular cell, allowing two cells from different organisms of the same species to combine.

Eukaryotic Cell Division: Meiosis

In sexually reproducing animals, it is usually necessary to reduce the genetic information before fertilization. Some plants can exist with too many copies of the genetic code, but in most organisms it is highly detrimental to have too many copies. Humans with even one extra copy of one chromosome can experience detrimental changes to their body. To counteract this, sexually reproducing organisms undergo a type of cell division known as meiosis. As before mitosis, the DNA and organelles are replicated. The process of meiosis contains two different cell divisions, which happen back-to-back. The first meiosis, meiosis I, separates homologous chromosomes. The homologous chromosomes present in a cell represent the two alleles of each gene an organism has. These alleles are recombined and separated, so the resulting daughter cells have only one allele for each gene, and no homologous pairs of chromosomes. The second division, meiosis II, separated the two copies of DNA, much like in mitosis. The end result of meiosis in one cell is 4 cells, each with only one copy of the genome, which is half the normal number.

Organisms typically package these cells into gametes, which can travel into the environment to find other gametes. When two gametes of the right type meet, one will fertilize the other and produce a zygote. The zygote is a single cell that will undergo mitosis to produce the millions of cells necessary for a large organism. Thus, most eukaryotes use both mitosis and meiosis, but at different stages of their lifecycle.

Cell Division Stages

Depending upon which type of cell division an organism uses, the stages can be slightly different.

Mitosis Stages

Mitosis starts with prophase in which the chromosome is condensed. The cell proceeds to metaphase where the chromosomes are aligned on the metaphase plate. Then the chromosomes are separated in anaphase and the cell's cytoplasm is pinched apart during telophase. Cytokinesis is the final process that breaks the cell membrane and divides the cell into two.

Meiosis Stages

The stages of meiosis are similar to mitosis, but the chromosomes act differently. Meiosis has two phases, which include two separate cell divisions without the DNA replicating between them. Meiosis I and meiosis II have the same 4 stages as mitosis: prophase, metaphase, anaphase, and telophase. Cytokinesis concludes both rounds of meiosis.

In prophase I, the chromosomes are condensed. In metaphase I, the chromosomes line up across from their homologous pairs. When they are separated in anaphase I and telophase I, there is only one form of each gene in each cell, known as a reduction division. Meiosis II proceeds in the same manner as mitosis, which sister chromatids dividing on the metaphase plate. By telophase II, there are 4 cells, each with half of the alleles as the parent cell and only a single copy of the genome. The cells can now become gametes and fuse together to create new organisms.

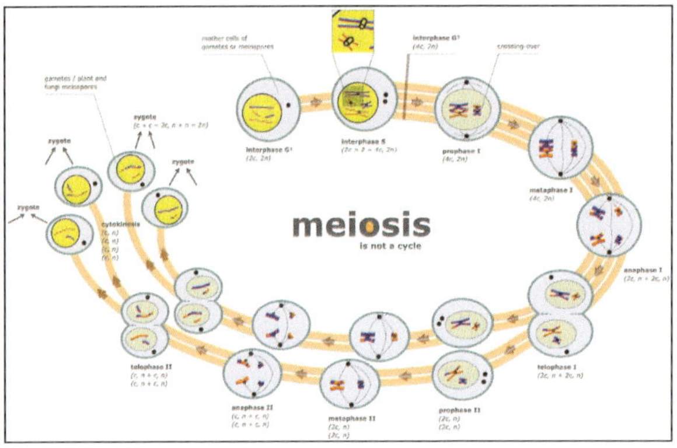

Cell Metabolism

Cell metabolism can be defined as a set of chemical process that helps an individual or an organism to respond to their surroundings. The cell metabolism is mainly involved in the extraction of energy that is used for many activities like cell growth, cells and tissue repair, etc. The cell extracts energy from breaking down the excess amount of carbohydrates, amino acids and lipids. This biological process is very much important for an organism, which is involved in obtaining chemical energy, synthesizing complex molecules and converting nutrient molecules into usable forms for further use . Without this process, a cell dies with lack of energy to carry out biological reactions. If a cell starts to die, finally an organism ends.

Types of Cell Metabolism

Cell metabolism can be generally classified into two major categories.

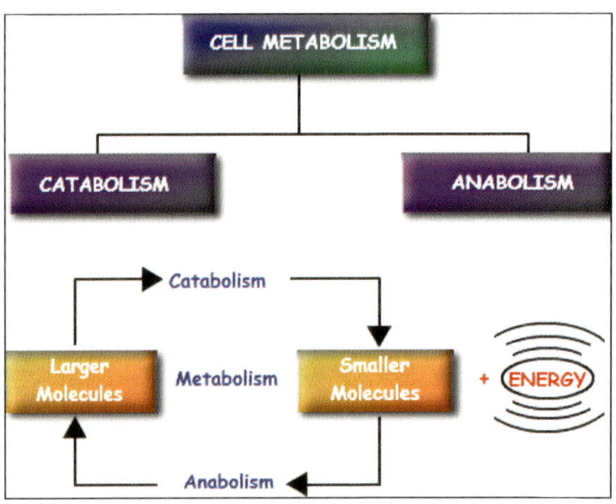

- Catabolism: In this reaction of metabolism, energy is obtained by the breakdown of the larger molecules into smaller ones. In this process the energy is released. Hydrolysis is an example of catabolism reaction.

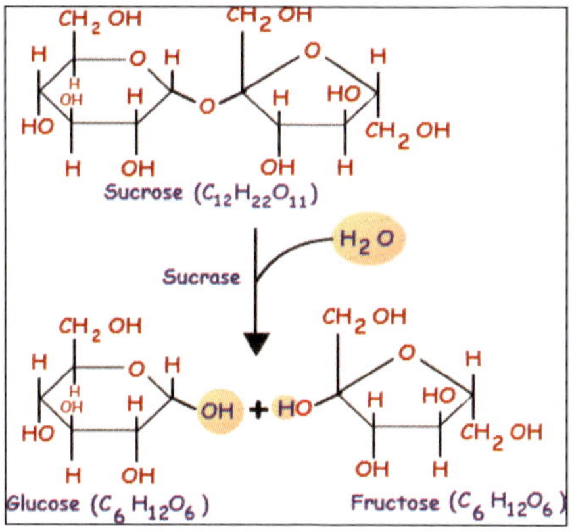

In the above reaction of hydrolysis, a water molecule is used.

- Anabolism: In this reaction of metabolism, cell synthesis all necessary compounds, which are required for cellular growth and repair. This necessary compound includes- DNA, RNA and proteins. This process requires energy to build larger molecules. Dehydration synthesis is an example of anabolism reaction.

Dehydration Synthesis -anabolism Reaction

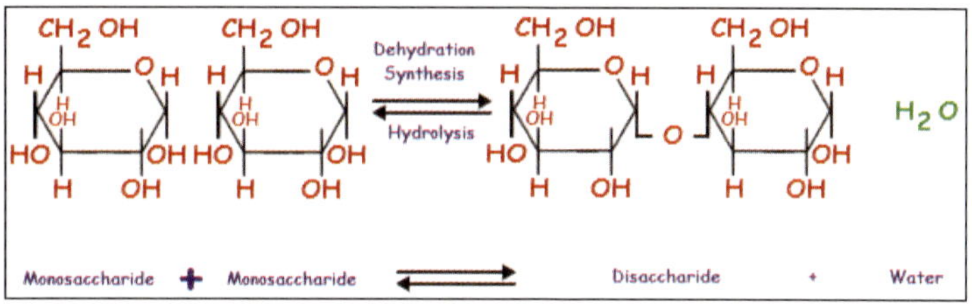

In the above reaction of dehydration synthesis, a water molecule is produced as a bi-product.

Plant Metabolism

Plant metabolism is the set of chemical reactions, which helps a plant to perform its own activities. Plants are the primary source of essential nutrients for billions of individuals globally. Plant metabolism is of two types.

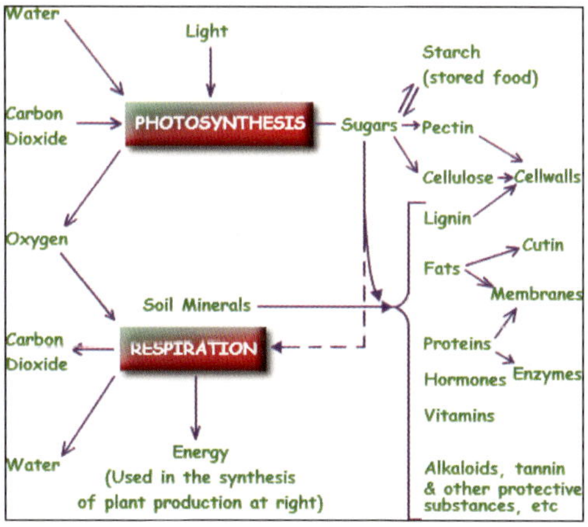

Types of Plant Metabolism

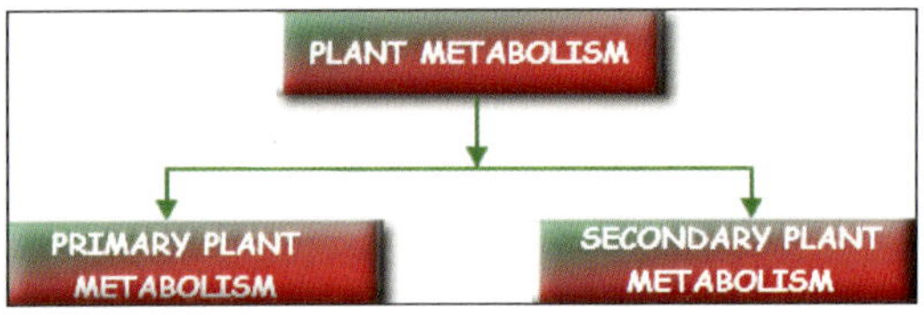

Primary Plant Metabolites

Primary metabolites are the compounds, which are commonly produced by all plants and are directly used by them for their growth and their development. Carbohydrates, proteins, nucleic acids and lipids are considered as the main primary metabolites.

- Functions of primary metabolic compounds.
- Polysaccharides are mainly used in providing support for plants and also stores energy for their later use.
- A protein builds the remaining biomass of living plant cells.
- A protein consists of polypeptides, which are made up of amino acids.
- Plants synthesize amino acids from the products of photosynthesis by using large amounts of energy in the form of ATP and NADPH.
- A protein stores energy in seeds and are used as a source of nutrition in the early development of seedlings. For example: Corn produces a storage protein called ZEIN.

Secondary Plant Metabolism

Secondary metabolites are the compounds, which are not required for normal growth and for the development of plant cells. The products of secondary metabolism are morphine, caffeine, nicotine, menthol and rubber. The products of secondary metabolism are the metabolism of chemicals, which are very rarely found in plants and do not have any specific role in plants functioning.

Functions of Secondary Metabolic Compounds.

- The product of secondary metabolism plays a vital role in interactions between plants and other organisms.
- They also play an ecological role in governing interactions between plants and other organisms.
- The product of secondary metabolism also helps in attracting pollinators, fruit and seed dispersers. These are because of their bright coloured pigments of flowers.
- Few toxic compounds, which are produced from secondary metabolism like nicotine, which helps the plant in protecting them from plant eating animals (herbivores) and also from the microbes.

A plant cell receives their energy from photosynthesis.

Photosynthesis

Plants depend on photosynthesis to obtain their energy. In plant cells, the presence of chlorophyll pigment helps to carry out the photosynthetic reaction. Photosynthesis is the process, which occurs in all green plants. The main role of photosynthesis is to trap solar energy and use this energy to drive the synthesis of carbohydrates from carbon dioxide. An equation for photosynthetic reaction is:

$$6CO_2 + 6H_2O + \text{sunlight à } C_6H_{12}O_6 + 6O_2$$

Photosynthesis is a biological process, which is carried out in two steps.

Light-dependent Reaction (or) Light Reaction

- In this biochemical process, light energy is converted into chemical energy and ATP molecules are produced.
- This process requires light and occurs in thylakoids in chloroplast cells.
- There is a loss of energy along the electron transport chain to recharge ATP from ADP.

Light-independent Reaction (or) Dark Reaction

- In this biochemical process, produced ATP molecules are used to convert them into simple sugars.
- This process does not require light and occurs in the stroma of chloroplast cells.
- In this process, ATP and NADPH are used to produce energy.
- Prepares glucose sugar from carbon dioxide and Hydrogen.

Calvin Cycle

The second step of photosynthesis, which takes place in the stroma of chloroplast cells and in the absence sunlight. In this biochemical process the carbon molecules from carbon dioxide (CO_2) are fixed into glucose ($C_6H_{12}O_2$).

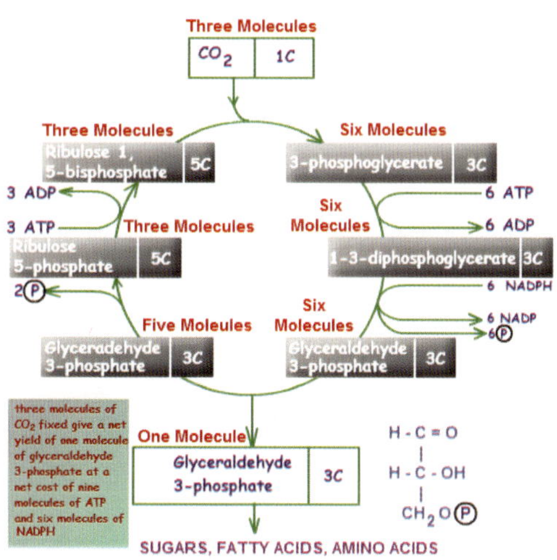

Calvin Cycle Reaction

Animal Metabolism

The chemical reactions, which occur in the cells of an animal body, are called as an animal's metabolism. In animal cells, mitochondria and the nucleus are the sites where an energy production reaction takes place. Cellular respiration is the type of metabolism that takes place in all animal cells.

Cellular Respiration

It is the process of respiration; energy is released by breaking down of food molecules in the presence of oxygen. An equation for cellular respiration is:

$$6O_2 + C_6H_{12}O_6 \rightarrow 6CO_2 + 6H_2O + \text{Energy}$$

Cellular respiration is carried out in 3 stages

Glycolysis – the first stage in cellular respiration.

- In this stage of cellular respiration, a series of enzyme catalyzes the reactions.
- In this process, glucose gets converted to pyruvic acid.
- Small numbers of ATPs molecules are formed.
- This process does not require oxygen and takes place in the cytoplasm of a cell.
- All living organisms use glycolysis.
- An equation for cellular respiration is

Glucose + 2ADP + 2P_i + 2 NAD^+à 2 Pyruvic acid + 2 NADH + 2ATP

Oxidation

- It harvests electrons and uses their energy to power ATP production.
- This process is a more powerful reaction and takes place only in the mitochondria of a cell.
- An atom that loses an electron gets oxidized. Oxygen is a common electron acceptor.

Anaerobic Respiration

- This process does not require oxygen and glucose is completely oxidized.
- Different types of electron acceptors are used in this process instead of oxygen (sulfur, or nitrate).

Comparing Photosynthesis (Plant metabolism) and Respiration (Animal metabolism).

	Photosynthesis (Plant metabolism)	Cellular Respiration (Animal metabolism)
Function	Energy Storage	Energy Release
	Chloroplasts	Mitochondria
Reactants	CO_2 and H_2O	$C_6H_{12}O_6$ and O_2

Products	$C_6H_{12}O_6$ and O_2	CO_2 and H_2O
Equation	$6CO_2 + 6H_2O \rightarrow C_6H_{12}O_6 + 6O_2$	$C_6H_{12}O_6 + 6O_2 \rightarrow 6CO_2 + 6H_2O$

Protein Biosynthesis

The protein synthesis which involves the translation of nucleotide base sequence of mRNA into the language of amino acid sequence may be divided into the following stages for the convenience of understanding.

1. Requirement of the components.
2. Activation of amino acids.
3. Protein synthesis proper.
4. Chaperones and protein folding.
5. Post-translational modifications.

Stage 1. Requirement of the Components:

The protein synthesis may be considered as a biochemical factory operating on the ribosomes. As a factory is dependent on the supply of raw materials to give a final product, the protein synthesis also requires many components.

a. Amino Acids:

Proteins are polymers of amino acids. Of the 20 amino acids found in protein structure, half of them (10) can be synthesized by man. About 10 essential amino acids have to be provided through the diet. Protein synthesis can occur only when all the amino acids needed for a particular protein are available.

If there is a deficiency in the dietary supply of any one of the essential amino acids, the translation stops. It is, therefore, necessary that a regular dietary supply of essential amino acids, in sufficient quantities, is maintained, as it is a prerequisite for protein synthesis. As regards prokaryotes, there is no requirement of amino acids, since all the 20 are synthesized from the inorganic components.

b. Ribosomes:

The functionally active ribosomes are the centres or factories for protein synthesis. Ribosomes may also be considered as workbenches of translation. Ribosomes are huge complex structures (70S for prokaryotes and 80S for eukaryotes) of proteins and ribosomal RNAs. Each ribosome consists of two subunits—one big and one small. The functional ribosome has two sites— A site and P site. Each site covers both the subunits.

A site is for binding of aminoacyl tRNA and P site is for binding peptidyl tRNA, during the course of translation. Some authors consider A site as acceptor site and P site as donor site. In case of eukaryotes, there is another site called exist site or E site. Thus, eukaryotes contain three sites (A, P and E) on the ribosomes.

The ribosomes are located in the cytosomal fraction of the cell. They are found in association with rough endoplasmic reticulum (RER) to form clusters RER—ribosomes, where the protein synthesis occurs. The term polyribosome (polysome) is used when several ribosomes simultaneously translate on a single mRNA (figure).

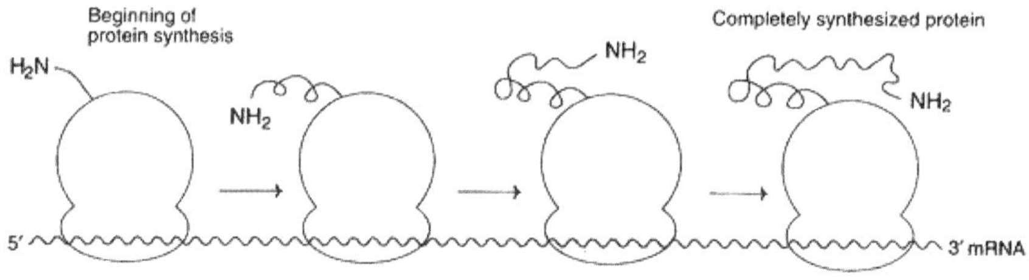

Figure: A polyribosome in protein synthesis

c. Messenger RNA (mRNA):

The specific information required for the synthesis of a given protein is present on the mRNA. The DNA has passed on the genetic information in the form of codons to mRNA to translate into a protein sequence.

d. Transfer RNAs (tRNAs):

They carry the amino acids, and hand them over to the growing peptide chain. The amino acid is covalently bound to tRNA at the 3′-end. Each tRNA has a three nucleotide base sequence—the anticodon, which is responsible to recognize the codon (complementary bases) of mRNA for protein synthesis. In man, there are about 50 different tRNAs whereas in bacteria around 40 tRNAs are found. Some amino acids (particularly those with multiple codons) have more than one tRNA.

e. Energy Sources:

Both ATP and GTP are required for the supply of energy in protein synthesis. Some of the reactions involve the breakdown of ATP or GTP, respectively, to AMP and GMP with the liberation of pyrophosphate. Each one of these reactions consumes two high energy phosphates (equivalent to 2 ATP).

f. Protein Factors:

The process of translation involves a number of protein factors. These are needed for initiation, elongation and termination of protein synthesis. The protein factors are more complex in eukaryotes compared to prokaryotes.

Stage 2. Activation of Amino Acids:

Amino acids are activated and attached to tRNAs in a two-step reaction. A group of enzymes—namely aminoacyl tRNA synthetases—are required for this process. These enzymes are highly specific for the amino acid and the corresponding tRNA. The amino acid is first attached to the enzyme utilizing ATP to form enzyme-AMP-amino acid complex. The amino acid is then transferred to the 3′ end of the tRNA to form aminoacyl tRNA (figure).

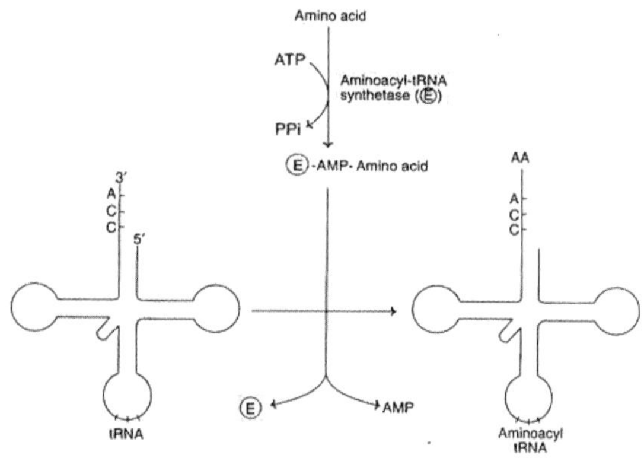

Figure: Formation fo aminocacyl tRNA (AA-Amino acid: E-Enzyme)

Stage 3. Protein Synthesis Proper:

The protein or polypeptide synthesis occurs on the ribosomes (rather polyribosomes). The mRNA is read in the direction and the polypeptide synthesis proceeds from N-terminal end to C-terminal end. Translation is directional and collinear with mRNA. The prokaryotic mRNAs are polycistronic, since a single mRNA has many coding regions that code for different polypeptides. In contrast, eukaryotic mRNA is monocistronic, since it codes for a single polypeptide.

In case of prokaryotes, translation commences before the transcription of the gene is completed. Thus, simultaneous transcription and translation are possible. This is not so in case of eukaryotic organisms since transcription occurs in the nucleus whereas translation takes place in the cytosol. Further, the primary transcript (hnRNA) formed from DNA has to undergo several modifications to generate functional mRNA.

Protein synthesis is comparatively simple in case of prokaryotes compared to eukaryotes. Further, many steps in eukaryotic translation were not understood for quite some time. For these reasons, majority of the textbooks earlier used to describe translation in prokaryotes in detail, and give most important and relevant information for eukaryotic translation. With the advances in molecular biology, the process of protein biosynthesis in eukaryotes is better understood now.

Translation in eukaryotes is briefly described here, along with some relevant features of prokaryotic protein biosynthesis. Translation proper is divided into three stages—initiation, elongation and termination.

Initiation of Translation

The initiation of translation in eukaryotes is complex, involving at least ten eukaryotic initiation factors (elFs). Some of the elFs contain multiple (3-8) subunits. The process of translation initiation can be divided into four steps (figure).

1. Ribosomal dissociation.
2. Formation of 43S pre-initiation complex.
3. Formation of 48S initiation complex.
4. Formation of 80S initiation complex.

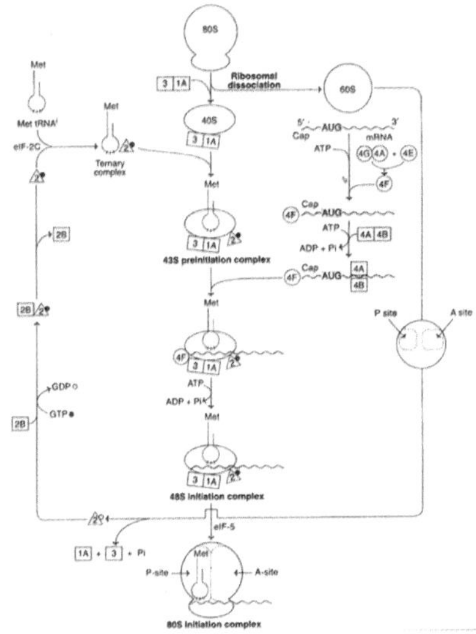

Figure: A diagrammatic representation of initation of protein biosynthesis (translation) in eukaryotic cells

Ribosomal dissociation

The 80S ribosome dissociates to form 40S and 60S subunits. Two initiating factors namely elF-3 and elF-1A bind to the newly formed 40S subunit, and thereby block its re-association with 60S subunit. For this reason, some workers name elF-3 as anti-association factor.

Formation of 43S pre-initiation complex

A ternary complex containing met-tRNA' and elF-2 bound to GTP attaches to 40S ribosomal subunit to form 43S pre-initiation complex. The presence of elF-3 and elF-1 A stabilizes this complex (Met-tRNA is specifically involved in binding to the initiation condon AUGs; hence the superscrip' is used in met-tRNA').

Formation of 48S initiation complex

The binding of mRNA to 43S pre-initiation complex results in the formation of 48S initiation complex through the intermediate 43S initiation complex. This, however, involves certain interactions between some of the elFs and activation of mRNA.

elF-4F complex is formed by the association of elF-4G, elF-4A with elF-4E. The so formed elF-4F (referred to as cap binding protein) binds to the cap of mRNA. Then elF-4A and elF-4B bind to mRNA and reduce its complex structure. This mRNA is then transferred to 43S complex. For the appropriate association of 43S pre-initiation complex with mRNA, energy has to be supplied by ATP.

Recognition of Initiation Codon

The ribosomal initiation complex scans the mRNA for the identification of appropriate initiation codon. 5'-AUG is the initiation codon and its recognition is facilitated by a specific sequence of nucleotides surrounding it. This marker sequence for the identification of AUG is called as Kozak consensus sequences. In case of prokaryotes the recognition sequence of initiation codon is referred to as Shine- Dalgarno sequence.

Formation of 80S initiation complex

48S initiation complex binds to 60S ribosomal subunit to form 80S initiation complex. The binding involves the hydrolysis of GTP (bound to elF-2). This step is facilitated by the involvement of elF-5. As the 80S complex is formed, the initiation factors bound to 48S initiation complex are released, and recycled. The activation of elF-2 requires elF-2B (also called as guanine nucleotide exchange factor) and GTP. The activated elF-2 (i.e. bound to GTP) requires elF-2C to form the ternary complex.

Regulation of Initiation

The elF-4F, a complex formed by the assembly of three initiation factors controls initiation, and thus the translation process. elF-4E, a component of elF-4F is primarily responsible for the recognition of mRNA cap. And this step is the rate-limiting in translation. elF-2 which is involved in the formation of 43S pre-initiation complex also controls protein biosynthesis to some extent.

Initiation of Translation in Prokaryotes

The formation of translation initiation complex in prokaryotes is less complicated compared to eukaryotes. The 30S ribosomal subunit is bound to initiation factor 3 (IF-3) and attached to ternary complex of IF-2, formyl met-tRNA and GTP.

Another initiation factor namely IF-I also participates in the formation of pre-initiation complex. The recognition of initiation codon AUG is done through Shine-Dalgarno sequence. A 50S ribosome unit is now bound with the 30S unit to produce 70S initiation complex in prokaryotes.

Elongation of Translation

Ribosomes elongate the polypeptide chain by a sequential addition of amino acids. The amino acid sequence is determined by the order of the codons in the specific mRNA. Elongation, a cyclic process involving certain elongation factors (EFs), may be divided into three steps (figure).

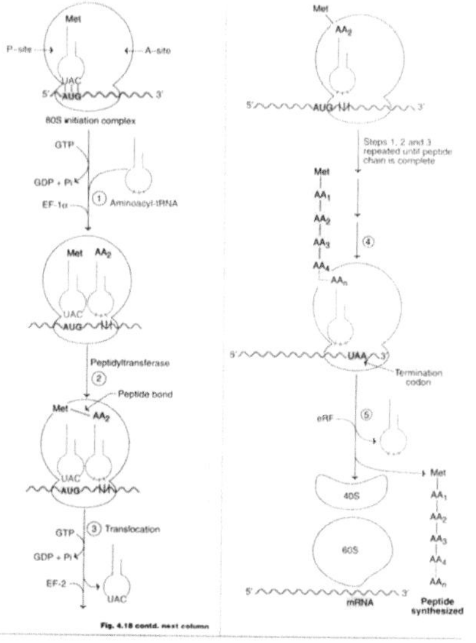

Figure: Protein biosynthesis—Elongation and termination

1. Binding of aminoacyl t-RNA to A-site.
2. Peptide bond formation.
3. Translocation.

Binding of Aminoacyl—tRNA to A-site:

The 80S initiation complex contains met-tRNA in the P-site, and the A-site is free. Another aminoacyl-tRNA is placed in the A-site. This requires proper codon recognition on the mRNA and the involvement of elongation factor 1a (EF-la) and supply of energy by GTP. As the aminoacyl-tRNA is placed in the A-site, EF-1a and GDP are recycled to bring another aminoacyl- tRNA.

Peptide Bond Formation

The enzyme peptidyltransferase catalyses the formation of peptide bond (Zigure). The activity of this enzyme lies on 28S RNA of 60S ribosomal subunit. It is therefore the rRNA (and not protein) referred to as ribozyme that catalyses the peptide bond formation. As the amino acid in the aminoacyl-tRNA is already activated, no additional energy is required for peptide bond formation.

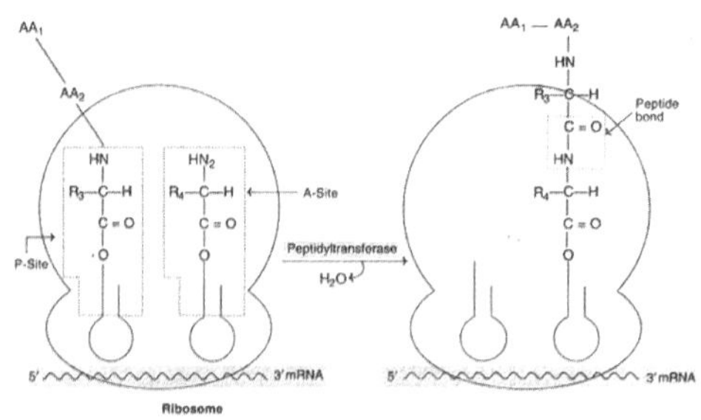

Figure: Formation of peptide bond in translation

The net result of peptide bond formation is the attachment of the growing peptide chain to the tRNA in the A-site.

Translocation

As the peptide bond formation occurs, the ribosome moves to the next codon of the mRNA (towards 3′-end). This process called translocation, basically involves the movement of growing peptide chain from A-site to P-site. Translocation requires EF-2 and GTP. GTP gets hydrolysed and supplies energy to move mRNA. EF-2 and GTP complex recycles for translocation.

In recent years, another site namely exist site (E-site) has been identified in eukaryotes. The deacylated tRNA moves into the E-site, from where it leaves the ribosome. In case of prokaryotes, the elongation factors are different, and they are EF-Tu, EF-Ts (in place of EF-1a) and EF-G (instead of EF-2).

Incorporation of amino acids:

It is estimated that about six amino acids per second are incorporated during the course of elongation of translation in eukaryotes. In case of prokaryotes, as many as 20 amino acids can be incorporated per second. Thus the process of protein/polypeptide synthesis in translation occurs with great speed and accuracy.

Termination of Translation

Termination is a simple process when compared to initiation and elongation. After several cycles of elongation, incorporating amino acids and the formation of the specific protein/polypeptide molecule, one of the stop or termination signals (UAA, UAG and UCA) terminates the growing polypeptide.

The termination codons which act as stop signals do not have specific tRNAs to bind. As the termination codon occupies the ribosomal A-site, the release factor namely eRF recognizes the stop signal. eRF-GTP complex, in association with the enzyme peptidyltransferase, cleaves the peptide bond between the polypeptide and the tRNA occupying P-site.

In this reaction, a water molecule, instead of an amino acid is added. This hydrolysis releases the protein and tRNA from the P-site. The 80S ribosome dissociates to form 40S and 60S subunits which are recycled. The mRNA is also released.

Inhibitors of Protein Synthesis

Translation is a complex process and it has become a favorite target for inhibition by antibiotics. Antibiotics are the substances produced by bacteria or fungi which inhibit the growth of other organisms. Majority of the antibiotics interfere with the bacterial protein synthesis and are harmless to higher organisms. This is due to the fact that the process of translation sufficiently differs between prokaryotes and eukaryotes. The action of a few important antibiotics on translation is described here.

- Streptomycin

Initiation of protein synthesis is inhibited by streptomycin. It causes misreading of mRNA and interferes with the normal pairing between codons and anticodons.

- Tetracycline

It inhibits the binding of aminoacyl tRNA to the ribosomal complex. In fact, tetracycline can also block eukaryotic protein synthesis. This, however, does not happen since eukaryotic cell membrane is not permeable to this drug.

- Puromycin

This has a structural resemblance to aminoacyl tRNA. Puromycin enters the A site and gets incorporated into the growing peptide chain and causes its release. This antibiotic prevents protein synthesis in both prokaryotes and eukaryotes.

- Chloramphenicol

It acts as a competitive inhibitor of the enzyme peptidyltransferase and thus interferes with elongation of peptide chain.

- Erythromycin

It inhibits translocation by binding with 50S subunit of bacterial ribosome.

- Diphtheria toxin

It prevents translocation in eukaryotic protein synthesis by inactivating elongation factor eEF2.

Stage 4. Chaperones and Protein Folding:

The three dimensional conformation of proteins is important for their biological functions. Some of the proteins can spontaneously generate the correct functionally active conformation e.g. denatured pancreatic ribonuclease. However, a vast majority of proteins can attain correct conformation, only through the assistance of certain proteins referred to as chaperones.

Chaperones are heat shock proteins (originally discovered in response to heat shock). They facilitate and favour the interactions on the polypeptide surfaces to finally give the specific conformation of a protein. Chaperones can reversibly bind to hydrophobic regions of unfolded proteins and folding intermediates.

They can stabilize intermediates, prevent formation of incorrect intermediates, and also prevent undesirable interactions with other proteins. All these activities of chaperones help the protein to attain compact and biologically active conformation.

Types of Chaperones

Chaperones are categorized into two major groups:

1. Hsp70 system:

This mainly consists of Hsp70 (70-kDa heat shock protein) and Hsp40 (40- kDa Hsp). These proteins can bind individually to the substrate (protein) and help in the correct formation of protein folding.

2. Chaperonin system:

This is a large oligomeric assembly which forms a structure into which the folded proteins are inserted. The chaperonin system mainly has Hsp60 and Hsp10 i.e. 60 kDa Hsp and 10 kDa Hsp. Chaperonins are required at a later part of the protein folding process, and often work in association with Hsp70 system.

Protein Misfolding and Diseases:

The failure of a protein to fold properly generally leads to its rapid degradation. Cystic fibrosis (CF) is a common autosomal recessive disease. Some cases of CF with mutations that result in altered protein (cystic fibrosis trans membrane conductance regulator or in short CFTR) have been reported. Mutated CFTR cannot fold properly, besides not being able to get glycosylated or transported. Therefore, CFTR gets degraded.

Certain neurological diseases which are due to cellular accumulation of aggregates of misfolded proteins or their partially degraded products have been identified. The term prions (proteinous infectious agents) is used to collectively represent them. Prions exhibit the characteristics of viral or microbial pathogens and have been implicated in many diseases, e.g. mad cow disease, Creutzfeldt- Jacob disease, Alzheimer's disease, Huntington's disease.

Stage 5. Post-Translational Modifications of Proteins:

The proteins synthesized in translation are, as such, not functional. Many changes take place in the polypeptides after the initiation of their synthesis or, most frequently, after the protein synthesis is completed. These modifications include protein folding, trimming by proteolytic degradation, intein splicing and covalent changes which are collectively known as post-translational modifications (figure).

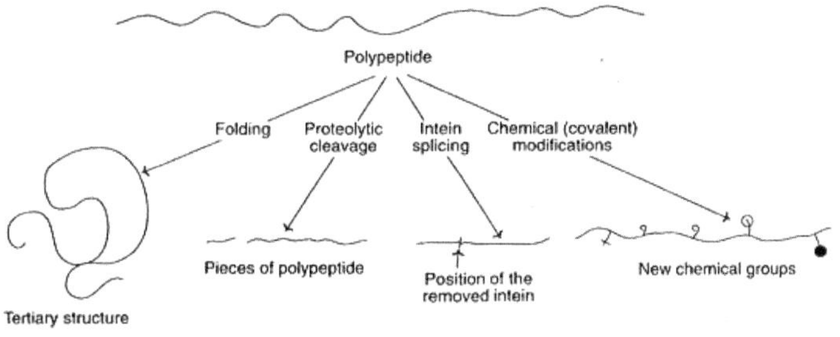

Figure: An outline of post-translational modifications of proteins

Proteolytic Degradation:

Many proteins are synthesized as the precursors which are much bigger in size than the functional proteins. Some portions of precursor molecules are removed by proteolysis to liberate active proteins. This process—commonly referred to as trimming- may occur in Golgi apparatus, secretory vesicles and, sometimes, after the secretion of proteins.

The formation of insulin from pre-pro-insulin, conversion of zymogens (inactive digestive enzymes e.g. trypsinogen) to the active enzymes are some examples of trimming. The synthesis of the proteins as inactive precursors and their later conversion into active form, may be, to protect the functional protein unit from the environmental insults.

Intein Splicing:

Inteins are intervening sequences in certain proteins. These are comparable to introns in mRNAs. Inteins have to be removed, and exteins ligated in the appropriate order for the protein to become active.

Covalent Modifications:

The proteins synthesized in translation are subjected to many covalent changes. By these modifications in the amino acids, the proteins may be converted to active form or inactive form.

Selected examples of covalent modifications are described below:

1. Phosphorylation:

The hydroxyl group containing amino acids of proteins, namely serine, threonine and tyrosine are subjected to phosphorylation. The phosphorylation may either increase or decrease the activity of the proteins. A group, of enzymes called protein kinases catalyses phosphorylation while protein phosphatases are responsible for dephosphorylation (removal of phosphate group). Many examples of enzymes that undergo phosphorylation or dephosphorylation are known in metabolisms.

2. Hydroxylation:

During the formation of collagen, the amino acids pro-line and lysine are respectively converted to hydroxyproline and hydroxylysine. This hydroxylation occurs in the endoplasmic reticulum and requires vitamin C.

3. Glycosylation:

The attachment of carbohydrate moiety is essential for some proteins to perform their functions. The complex carbohydrate moiety is attached to the amino acids, serine and threonine (O-linked) or to asparagine (N-linked), leading to the synthesis of glycoproteins. Vitamin K dependent carboxylation of glutamic acid residues in certain clotting factors is also a post-translational modification.

Table: Selected examples of post-translational modifications of proteins through their amino acids

Amino acid	*Post-translational modification(s)*
Amino-terminal amino acid	Glycosylation, acetylation, myristoylation, formylation.
Carboxy terminal amino acid	Methylation, ADP-ribosylation
Arginine	Methylation
Aspartic acid	Phosphorylation, hydroxylation
Cysteine (—SH)	Cystine (—S—S—) formation, selenocysteine formation, glycosylation.
Glutamic acid	Methylation, γ-carboxylation.
Histidine	Methylation, phosphorylation.
Lysine	Acetylation, methylation, hydroxylation, biotinylation.
Methionine	Sulfoxide formation.
Phenylalanine	Glycosylation, hydroxylation.
Proline	Hydroxylation, glycosylation.
Serine	Phosphorylation, glycosylation.
Threonine	Phosphorylation, methylation glycosylation.
Tryptophan	Hydroxylation.
Tyrosine	Hydroxylation, phosphorylation, sulfonylation, iodination.

Motility

Motility is a biological term which refers to the ability to move spontaneously and actively, consuming energy in the process. Most animals are motile but the term applies to single-celled and simple multicellular organisms, as well as to some mechanisms of fluid flow in multicellular organs, in addition to animal locomotion. Motile marine animals are commonly called free-swimming. The opposite of motility is sessility. Motility may also refer to an organism's ability to move food through its digestive tract, i.e., peristaltics (gut motility, intestinal motility, etc.).

Cells exhibit a wide range of movement. These movements include migration of cells along a surface or through a tissue, or movement of components within cells. Specific examples of cell motility include:

- Movement of cells from one location in an embryo to another during embryonic development.
- Migration of cells into a wound during wound healing.
- Contraction of a muscle cell that is the fundamental process responsible for muscle contraction.
- Separation of a cell into two daughter cells during cell division.
- Movement of membrane-bound vesicles into cells during phagocytosis or endocytosis.
- Movement of membrane-bound vesicle from the cell interior to the cell surface during secretion.
- Movement of chromosomes during mitosis.

The first four bulleted points are examples of cell movement, while the last three bulleted points are examples of "intracellular motility." All of these movements have in common the fact that they are mediated by filamentous structures in the cell called the cytoskeleton and are powered by molecular motors that move along these filamentous structures. The simplest example is the movement of membrane-bound vesicles. These vesicles bind to a molecular motor just like a boxcar attaches to a railroad locomotive. The vesicle represents the cargo and the molecular motor represents the locomotive. The molecular motor then moves along the cytoskeletal filament as a locomotive moves along a railroad track. Most forms of intracellular movement occur using this mechanism.

A second mechanism is called contraction. This mechanism is responsible for contraction of muscle cells and the separation of daughter cells during cell division. Contraction works through the action of molecular motors pulling on the cytoskeletal filaments, drawing them toward each other. A third mechanism involves the rapid polyermization of the cytoskeleton. In this case the filamentous structures (usually the microfilament cytoskeleton) extend by the addition of subunits to the end. This growth of filaments then pushes out the membrane. This mechanism is responsible for protrusion of the front end of migrating cells.

Moving cells exhibit a special kind of directional movement called "chemotaxis." This mechanism accounts for the ability of cells to migrate in a specific direction. During chemotaxis, cells move in response to an external signal, most frequently a small molecule or short peptide, called a chemoattractant. Cells sense the concentration of the chemical and move in the direction of increasing concentration of the signal. This directional movement is responsible for much of the cell migration required for tissue formation and for wound healing. Wounded cells, for instance, release chemoattractants that attract immune system cells called macrophages and connective tissue cells called fibroblast .

For most eukaryotic cells , the process of cell movement occurs in several coordinated steps. First, cells extend a structure called a pseudopod ("false foot") using the polymerization mechanism.

Next the pseudopod makes an attachment to the surface along which the cell is moving. This establishes the new front of the cell. A contraction-mediated process provides the force that moves the rest of the cell toward the front and leads to detachment of the trailing end of the cell. When cells are exhibiting chemotaxis they extend pseudopods in several directions, but only make the attachment to the surface in the direction of the highest concentration of the chemical signal.

In contrast, bacterial cells move by the action of an amazing rotary motor called a bacterial flagellum. This flagellum spins like a propeller propelling the cell forward. Bacteria undergo chemotaxis by a process called tumble and run. In this process, a bacterial cells tumbles end over end and then "runs," moving in a single (random) direction for a defined period of time. At the end of that period the cell stops, tumbles again, and measures the concentration of the chemoattractant. If the concentration of chemoattractant is higher than at the last sampling the cell runs for an increased distance. If the chemoattractant concentration is lower, the run distance is shortened and the cell tumbles more frequently. Through this biased process cells preferentially migrate in the direction of higher chemoattractants.

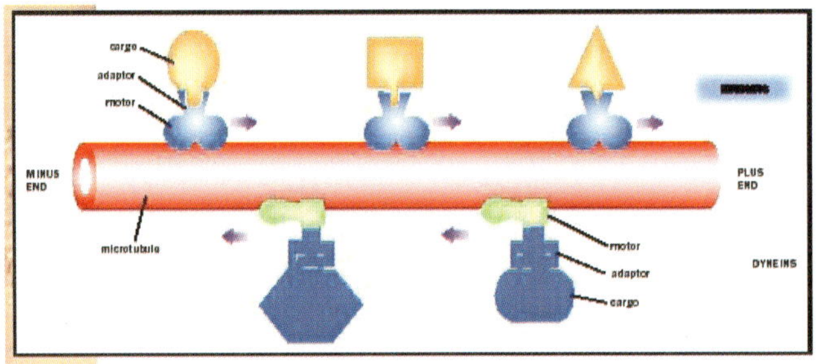

The motor proteins that move along microtubules. Kinesins move toward the plus end, whereas dyneins move toward the minus end.

References

- Structure-of-plant-cell-explained-with-diagram-2511: biologydiscussion.com, Retrieved 25 May 2018
- Cell-structure-definitions-5043056: sciencing.com, Retrieved 25 May 2018
- Six-main-cell-functions-6891800: sciencing.com, Retrieved 24 March 2018
- Major-functions-cell-112921: livestrong.com, Retrieved 09 July 2018
- Prokaryotes-meaning-373369: thoughtco.com, Retrieved 20 April 2018
- Cell-Motility: biologyreference.com, Retrieved 10 July 2018

3

Genetics

Genetics is concerned with the study of genes, variation and heredity in organisms. To develop an understanding of biology, it is vital to delve into the fundamentals of DNA, RNA, chromosomes, genetic inheritance, genetic drift as well as the process of meiosis and mitosis, which have been extensively covered in this chapter.

Genes

Genes is the units of heredity in living organisms which are encoded in an organism's genetic material (DNA). They exert a central influence on the organism's physical aspects and are passed on to succeeding generations through the reproduction process. Genetic material can also be passed between unrelated individuals on viruses or through the process of transfection used in genetic engineering.

Common usage of the word "gene" reflects its meaning in molecular biology, namely the segments of DNA that cells transcribe into either RNA that is translated into proteins (DNA=>RNA=>protein) or RNA used for direct purposes (DNA=>RNA). The Sequence Ontology project, a consortium of several centers of genomic studies, defines a gene as: "A locatable region of genomic sequence, corresponding to a unit of inheritance, which is associated with regulatory regions, transcribed regions, and/or other functional sequence regions." The definition reflects the full complexity that has come to be associated with the term gene.

Genes encode the information necessary for constructing the multitude of proteins and RNA units needed to maintain an organism's existence, growth, action, and multiplication. Each gene that serves as the first step in protein formation is a region of DNA comprising a mixture of some sections (exons) that code for proteins, others (introns) that have no apparent function, and still others that define the beginning and end of the gene or the conditions in which the gene will be expressed or not expressed.

Although the human genome comprises roughly 25,000 genes carrying codes for proteins, each human cell has the potential of making about 100,000 different proteins. Further complexity lies in the additional 10,000 or so genes used for making RNA that directly serves such cellular functions as structure, catalysis, and regulation of gene expression. The proteins and RNAs all share in the tasks of maintaining the cell, one of which is the continual fine tuning of the exact selection of genes being expressed according to the cell's function and its continually changing environment. The ongoing discovery of so much functional RNA in the cell, much of it related to the expression of genes, is taken be some as a sign that RNA may deserve a co-equal billing with DNA in terms of overall contribution to the cellular function.

Genes are of central importance to the physical aspect of a living organism: A person's eye color, the breed of a dog, the gender of a horse. Mouse DNA yields a mouse, not an elephant. However, the impact of genes is sometimes extrapolated to the view that genes control everything about human lives and destiny. This is the concept of genetic determinism whereby human behavior, intelligence, emotions and attitudes, and health are fixed by genetic makeup and thus unchangeable. Such a misconception has at times been used as a base for explaining away racial prejudices, addictions, and criminal behavior, and seeking solutions to social problems by turning to genetic engineering as the ultimate solution.

The more balanced and generally recognized view is that biological contributions to solving social problems must be sought through a biology that takes into account the influence of social and cultural factors in human physical development and behavior.

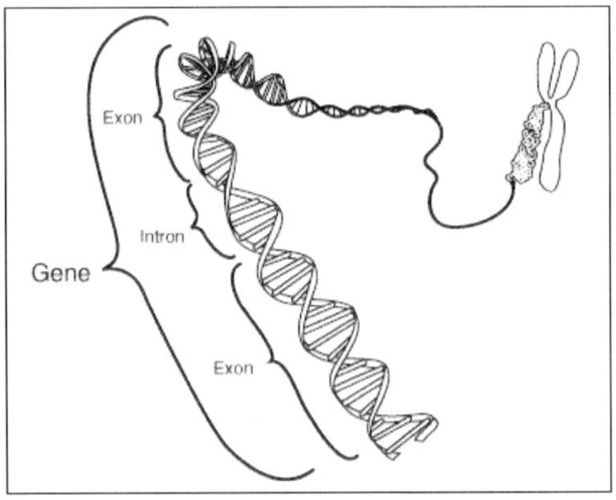

Diagram showing a greatly simplified gene in relation to the double helix structure of DNA with intron (non-coding) and exon (coding) portions labeled. If the gene is active in a particular cell, it will be transcribed into RNA that serves either a direct function in the cell or as a template for making a protein. In eukaryotic organisms, making protein from RNA entails excising the intron regions and splicing together some of the exon regions to make the template for one or more different proteins depending on which of the RNA exon regions are spliced together.

In molecular biology, a gene is considered to comprise both a coding sequence—the region of DNA (or RNA, in the case of some viruses) that determines the structure of a protein—and a regulatory sequence—the region of DNA that controls when and where the protein will be produced. The genetic code determines how the coding DNA sequence is converted into a protein sequence (via transcription and translation). The genetic code is essentially the same for all known life, from bacteria to humans.

Through the proteins they encode, genes govern the cells in which they reside. In multicellular organisms, much of the development of the individual, as well as the day-to-day functions of the cells, is tied to genes. The genes' protein products fulfill roles ranging from mechanical support of the cell structure to the transportation and manufacture of other molecules and the regulation of other proteins' activities.

Due to rare, spontaneous changes (e.g. in DNA replication) mutations in the sequence of a gene may arise. If these mutations occur in the germ line cells, they may be passed on to the organism's offspring. Once propagated to the next generation, this mutation may lead to variations within the population of a species. Variants of a single gene are known as alleles, and differences in alleles may give rise to differences in traits, for example, eye color. A gene's most common allele is called the wild type allele, and rare alleles are called mutants.

The genotype of an individual organism is its specific genetic makeup (the specific genome). The phenotype of an individual organism is determined to some extent by the genotype, or by the identity of the alleles that an individual carries at one or more positions on the chromosomes. A phenotype is either the organism's total physical appearance and constitution or a specific manifestation of a trait, such as size, eye color, or behavior that varies between individuals. Many phenotypes are determined by multiple genes and influenced by environmental factors.

In most cases, RNA is an intermediate product in the process of manufacturing proteins from genes. However, for some gene sequences, the RNA molecules are the actual functional agents. For example, RNAs known as ribozymes are capable of enzymatic function, and small interfering RNAs have a regulatory role. The DNA sequences from which such RNAs are transcribed are known as genes for non-coding RNA or RNA.

Most living organisms carry their genes as, and transmit them to offspring as, DNA, but some viruses carry only RNA. Because these viruses use RNA, their cellular hosts may synthesize the viral proteins as soon as they are infected and without the delay in waiting for transcription. On the other hand, RNA retroviruses, such as HIV, require the reverse transcription of their genome from RNA into DNA before their proteins can be synthesized.

In common speech, "gene" is often used to refer to the hereditary cause of a trait, disease, or condition—as in "the gene for obesity." Speaking more precisely, a biologist might refer to an allele or a mutation that "has been implicated in" or "is associated with" obesity. This is because biologists know that many factors other than genes decide whether a person is obese or not: Eating habits, exercise, prenatal environment, upbringing, culture, and the availability of food, for example.

Moreover, it is highly unlikely that variations within a single gene, or single genetic locus, would fully determine an individual's genetic predisposition for obesity. Rather, the norm with regard to many and perhaps most ("complex" or "multi-factoral") traits is that they reflect the combined effects of several factors including inheritance, interplay between genes and environment, and the combined influence of many genes. The term phenotype refers to the physical characteristics result from the interplay of all of these factors.

Typical Numbers of Genes in an Organism

This table gives typical numbers of genes and genome size for some organisms. Estimates of the number of genes in an organism are somewhat controversial because they depend on the discovery of genes, and no techniques currently exist to prove that a DNA sequence contains no gene. (In early genetics, genes could be identified only if there were mutations, or alleles.) Nonetheless, estimates are made based on current knowledge.

organism	genes	base pairs
Plant	<50,000	$<10^{11}$
Human, mouse or rat	25,000	3×10^{9}

Fruit Fly	13,767	1.3×10^8
Honey bee	15,000	3×10^8
Worm	19,000	9.7×10^7
Fungus	6,000	1.3×10^7
Bacterium	500–6,000	5×10^5–10^7
Mycoplasma genitalium	500	580,000
DNA virus	10–900	5,000–800,000
RNA virus	1–25	1,000–23,000
Viroid	0–1	~500

Human Gene Nomenclature

For each known human gene, the HUGO Gene Nomenclature Committee (HGNC) approves a gene name and symbol and stores all approved symbols in the HGNC Database. Each symbol is unique and each gene is given only one symbol. This protocol greatly facilitates clear and precise gene identifications in communications and in electronic data retrieval from publications. By convention, symbols for the different genes within a gene family all share a certain parallelism of construction. The symbols for human genes can also be applied to congruent genes in other species, such as the mouse.

Evolution and Genes

Broadly defined, evolution is any heritable change in a population of organisms over time.

The changes in populations that are considered evolutionary are those that are inheritable via the genetic material from one generation to another. As such, evolution can also be defined in terms of allele frequency, with allele being alternative forms of a gene, such as an allele for blue eye color versus brown eye color. Two important and popular evolutionary theories that address the pattern and process of evolution are the theory of descent with modification and the theory of natural selection.

The theory of descent with modification, or the "theory of common descent" deals with the pattern of evolution and essentially postulates that all organisms have descended from common ancestors by a continuous process of branching. The theory of modification through natural selection, or the "theory of natural selection," deals with mechanisms and causal relationships, and offers one explanation for how evolution might have occurred—the process by which evolution took place to arrive at the pattern.

According to the modern evolutionary synthesis, which integrated Charles Darwin's theory of evolution by natural selection with Gregor Mendel's theory of genetics as the basis for biological inheritance and mathematical population genetics, evolution consists primarily of changes in the frequencies of alleles between one generation and another as a result of natural selection. Natural selection has traditionally been viewed as acting on individual organisms, but has also been seen as working on groups of organisms.

An alternative model, the gene-centered view of evolution, sees natural selection as working on the level of genes.

Gene-centered View of Evolution

The gene-centered view of evolution, gene selection theory, or selfish gene theory, holds that natural selection acts through differential survival of competing genes, increasing the frequency of those alleles whose phenotypic effects successfully promote their own propagation. According to this theory, adaptations are the phenotypic effects through which genes achieve their propagation.

The view of the gene as the unit of selection was mainly developed in the books Adaptation and Natural Selection, by George C. Williams, and also in The Selfish Gene and The Extended Phenotype, both by Richard Dawkins.

Essentially, this view notes that the genes in existence today are those that have reproduced successfully in the past. Often, many individual organisms share a gene; thus, the death of an individual need not mean the extinction of the gene. Indeed, if the sacrifice of one individual enhances the survivability of other individuals with the same gene, the death of an individual may enhance the overall survival of the gene. This is the basis of the selfish gene view, popularized by Richard Dawkins. He points out in his book, The Selfish Gene, that to be successful, genes need have no other "purpose" than to propagate themselves, even at the expense of their host organism's welfare. A human that behaved in such a way would be described as "selfish," although ironically a selfish gene may promote altruistic behaviors. According to Dawkins, the possibly disappointing answer to the question "what is the meaning of life?" may be "the survival and perpetuation of ribonucleic acids and their associated proteins."

However, a number of prominent evolutionists, including Ernst Mayr and Stephen Jay Gould, who do recognize selection at levels other than the individual, nonetheless strongly reject the selfish gene theory. Mayr states that "the reductionist thesis that the gene is the object of selection" is "invalid." Gould calls the theory a "conceptual error" that sidetracked the profession, and "inspired both a fervent following of a quasi-religious nature" and "strong opposition from many evolutionists."

Chemistry and Function of Genes

A DNA molecule or strand comprises four kinds of sequentially linked nucleotides, which together constitute the genetic alphabet. A sequence of three consecutive nucleotides, called a codon, is the protein-coding vocabulary. The sequence of codons in a gene specifies the amino acid sequence of the protein it encodes.

In most eukaryotic species, very little of the DNA in the genome actually encodes proteins, and the genes may be separated by vast sequences of so-called "junk DNA." Moreover, the genes are often fragmented internally by non-coding sequences called introns, which can be many times longer than the coding sequence. Introns are removed on the heels of transcription by splicing. In the primary molecular sense, however, they represent parts of a gene.

All the genes and intervening DNA together make up the genome of an organism, which in many species is divided among several chromosomes and typically present in two or more copies. The location (or locus) of a gene and the chromosome on which it is situated is, in a sense, arbitrary. Genes that appear together on the chromosomes of one species, such as humans, may appear on separate chromosomes in another species, such as mice. Two genes positioned near one another on a chromosome may encode proteins that figure in the same cellular process or in completely unrelated processes. As an example of the former, many of the genes involved in spermatogenesis reside together on the Y chromosome.

Many species carry more than one copy of their genome within each of their somatic cells. These organisms are called diploid if they have two copies, or polyploid if they have more than two copies. In such organisms, the copies are practically never identical. With respect to each gene, the copies that an individual possesses are liable to be distinct alleles, which may act synergistically or antagonistically to generate a trait or phenotype. The ways that gene copies interact are explained by chemical dominance relationships.

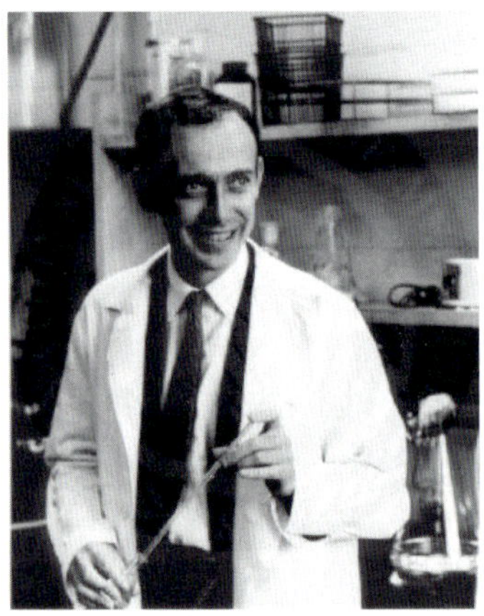

James Watson, one of discoverers of the structure of the DNA molecule.

Expression of Molecular Genes

For various reasons, the relationship between a DNA strand and a phenotype trait is not direct. The same DNA strand in two different individuals may result in different traits because of the effect of other DNA strands or the environment.

- The DNA strand is expressed into a trait only if it is transcribed to RNA. Because the transcription starts from a specific base-pair sequence (a promoter) and stops at another (a terminator), the DNA strand needs to be correctly placed between the two. If not, it is considered junk DNA, and is not expressed.
- Cells regulate the activity of genes in part by increasing or decreasing their rate of transcription. Over the short term, this regulation occurs through the binding or unbinding of proteins, known as transcription factors, to specific non-coding DNA sequences called regulatory elements. Therefore, to be expressed, the DNA strand needs to be properly regulated by other DNA strands.
- The DNA strand may also be silenced through DNA methylation or by chemical changes to the protein components of chromosomes.

- The RNA is often edited before its translation into a protein. Eukaryotic cells splice the transcripts of a gene by keeping the exons and removing the introns. Therefore, the DNA strand needs to be in an exon to be expressed. Because of the complexity of the splicing process, one transcribed RNA may be spliced in alternate ways to produce not one, but a variety of proteins (alternative splicing) from one pre-mRNA (mRNA transcript at pre-splicing stage). Prokaryotes produce a similar effect by shifting reading frames (the three ways the mRNA can be read by grouping the nucleotides into sets of three, as codons) during translation.
- The translation of RNA into a protein also starts with a specific start and stop sequence.
- Once produced, the protein interacts with the many other proteins in the cell, according to the cell metabolism. This interaction finally produces the trait.

This complex process helps explain the different meanings of "gene":

- nucleotide sequence in a DNA strand;
- or the transcribed RNA, prior to splicing;
- or the transcribed RNA after splicing, i.e. without the introns

The latter meaning of gene is the result of a more "material entity" than the first one.

Mutations and Evolution

Just as there are many factors influencing the expression of a particular DNA strand, there are many ways to have genetic mutations.

For example, natural variations within regulatory sequences appear to underlie many of the heritable characteristics seen in organisms. The influence of such variations on the trajectory of evolution may be as large as or larger than variation in sequences that encode proteins. Thus, though regulatory elements are often distinguished from genes in molecular biology, in effect they satisfy the shared and historical sense of the word. Indeed, a breeder or geneticist, in following the inheritance pattern of a trait, has no immediate way of knowing whether this pattern arises from coding sequences or regulatory sequences. Typically, he or she will simply attribute it to variations within a gene.

Errors during DNA replication may lead to the duplication of a gene, which may diverge over time. Though the two sequences may remain the same, or be only slightly altered, they are typically regarded as separate genes (i.e. not as alleles of the same gene). The same is true when duplicate sequences appear in different species. Yet, though the alleles of a gene differ in sequence, nevertheless they are regarded as a single gene (occupying a single locus).

The Shifting Locus of Biological Centrality

When the Human Genome Project began in 1990 scientists' estimates of the number of genes they would find was roughly 100,000-150,000, largely because of the number of different kinds of proteins found in the body and the assumption that one gene coded for one protein. By the end of the project in 2003, the estimate was 20,000 to 25,000 genes that coded for proteins, which was taken to mean that many genes must be coding for two, three, or four, or perhaps more different kinds of proteins.

This marked the beginning of a shift from the sense that DNA and the genes carried on it exercise singular influence and control in shaping the physical potentials of an individual. If one gene makes more than one protein, then the mechanism deciding which protein is produced from a given gene would be critical to the shaping of the individual's physical potentials. Of comparable importance to the question of the centrality of the genes is that of how a given human cell selects a subset of the 20-25,000 genes that will ultimately yield the 10,000 or so proteins that the cell needs out of the roughly 100,000 proteins available to it.

After the several decades in which DNA and the genes carried on it have been widely treated as the "stars" of the cellular world, new candidates are challenging for coequal or even perhaps primary recognition in terms of central importance to the cellular function. One, tied to the RNA World model of the origins of life, notes the growing number of identified types of non-coding functional RNA, many of which play a role in the regulation of gene expression. In this view, DNA is a passive, unchanging repository of information, whereas RNA is the active information agent even influencing which segments of DNA are expressed. This view suggests that RNA must deserve at least a co-equal place with DNA as a factor influencing an organism's physiology and psychology.

The second view shifts the focus completely away from the cell nucleus, DNA, and RNA. It notes that cells alter the selection of genes they express according to environmental influences they experience, and further that cells experience the environment through the mediation of the protective cell membrane and the thousands of proteins floating in it. With membrane proteins being sensitive to both magnetic and electromagnetic signals, the cells become object partners to their immediate environment(epigenetic factors), which includes influences from thoughts and emotions of the human mind responding to the environment (such as the adrenaline rush when a person wakes up in a burning house). In this view, mind becomes the intermediate third actor in the traditional dichotomy of nature (genes) or nurture (environment).

DNA

DNA, or deoxyribonucleic acid, is the hereditary material in humans and almost all other organisms. Nearly every cell in a person's body has the same DNA. Most DNA is located in the cell nucleus (where it is called nuclear DNA), but a small amount of DNA can also be found in the mitochondria (where it is called mitochondrial DNA or mtDNA). Mitochondria are structures within cells that convert the energy from food into a form that cells can use.

The information in DNA is stored as a code made up of four chemical bases: adenine (A), guanine (G), cytosine (C), and thymine (T). Human DNA consists of about 3 billion bases, and more than 99 percent of those bases are the same in all people. The order, or sequence, of these bases determines the information available for building and maintaining an organism, similar to the way in which letters of the alphabet appear in a certain order to form words and sentences.

DNA bases pair up with each other, A with T and C with G, to form units called base pairs. Each base is also attached to a sugar molecule and a phosphate molecule. Together, a base, sugar, and phosphate are called a nucleotide. Nucleotides are arranged in two long strands that form a spiral called a double helix. The structure of the double helix is somewhat like a ladder, with the base pairs forming the ladder's rungs and the sugar and phosphate molecules forming the vertical side-pieces of the ladder.

An important property of DNA is that it can replicate, or make copies of itself. Each strand of DNA in the double helix can serve as a pattern for duplicating the sequence of bases. This is critical when cells divide because each new cell needs to have an exact copy of the DNA present in the old cell.

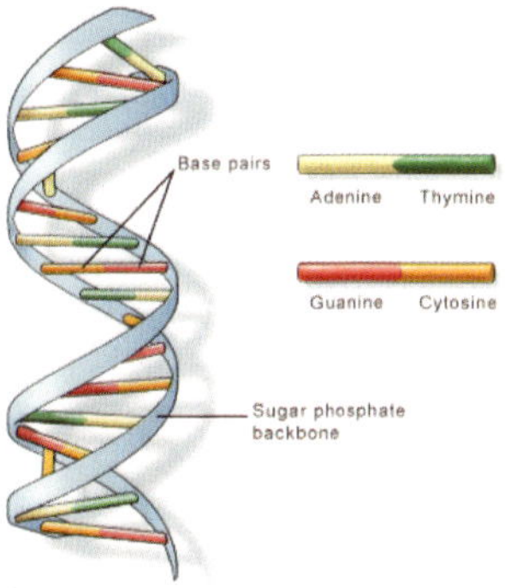

Figure: DNA is a double helix formed by base pairs attached to a sugar-phosphate backbone.

Types of DNA

DNA duplex model proposed by Watson and Crick is right handed spiral and is called B-DNA (Balanced DNA). In the model the base pairs lie at nearly right angles to the axis of helix. Another right handed duplex model is A-DNA (Alternate DNA). Here, a single turn of helix has 11 base pairs.

The base pairs lie 20° away from perpendicular to the axis. C-DNA has 9 base pairs per turn of spiral while in D-DNA the number is only 8 base pairs. Both are right handed. Z-DNA (Zigzag DNA) is left-handed double helix with zigzag back-bone, alternate purine and pyrimidine bases, single turn of 45 A length with 12 base pairs and a single groove.

B-DNA is more hydrated and most frequently found DNA in living cells. It is physiologically and biologically active form. However, it can get changed into other forms. Right handed DNA is known to change temporarily into the left handed form at least for a short distance. Such changes may cause changes in gene expression.

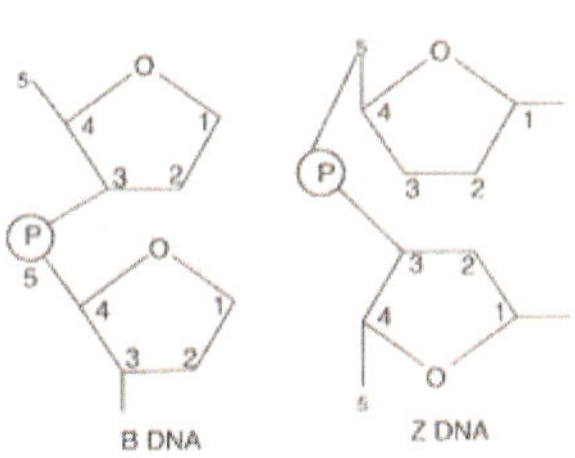

Figure: Orientation of adjacent sugar molecules in B and Z DNA.

Circular and Linear DNA:

- In many prokaryotes the two ends of a DNA duplex are covalently linked to form circular DNA. Circular DNA is naked, that is, without association with histone proteins, though polyamines do occur. In linear DNA the two ends are free. It is found in eukaryotic nuclei where it is associated with histone proteins.

- Linear DNA, without association with histone proteins, also occurs in some prokaryotes, e.g., Mycoplasma. In semi-autonomous cell organelles (mitochondria, plastids) DNA is circular, less commonly linear. It is always naked.

Chargaff's Rules

Chargaff made observations on the bases and other components of DNA. These observations or generalizations are called Chargaff's base equivalence rule.

i) Purine and pyrimidine base pairs are in equal amount, that is, adenine + guanine = thymine + cytosine. [A + G] = [T + C], i.e., [A+G] / [T+C] = 1

ii) Molar amount of adenine is always equal to the molar amount of thymine. Similarly, molar concentration of guanine is equalled by molar concentration of cytosine.

[A] = [T], i.e., [A] / [T] = 1; [G] = [C], i.e., [G] / [C] = 1

iii) Sugar deoxyribose and phosphate occur in equimolar proportions.

iv) A-T base pairs are rarely equal to C—G base pairs.

v) The ratio of [A+T] / [G+C] is variable but constant for a species (Table). It can be used to identify the source of DNA. The ratio is low in primitive organisms and higher in advanced ones.

Table. Base Composition of DNA from Various Sources:

Species	A	G	C	T	A+T/C+G
1. Man	30.4	19.0	19.9	30.1	1.55
2. Calf	29.0	21.2	21.2	28.5	1.35
3. Wheat germ	28.1	21.8	22.7	27.4	1.25
4. Pea	30.8	19.2	18.5	30.5	1.62
5. Euglena	22.6	27.7	25.8	24.4	0.88
6 Escherichia coli	24.7	26.0	25.7	23.6	0.93

Structure of DNA

DNA or deoxyribonucleic acid is a helically twisted double chain polydeoxyribonucleotide macromolecule which constitutes the genetic material of all organisms with the exception of rhinoviruses. In prokaryotes it occurs in nucleoid and plasmids. This DNA is usually circular. In eukaryotes, most of the DNA is found in chromatin of nucleus.

It is linear. Smaller quantities of circular, double stranded DNA are found in mitochondria and plastids (organelle DNA). Small sized DNAs occur in viruses, ϕ x 174 bacteriophage has 5386 nucleotides. Bacteriophage lambda (Phage X) possesses 48502 base pairs (bp) while number of base pairs in Escherichia coli is 4.6×10^6. A single genome (haploid set of 23 chromosomes) has about 3.3×10^9 bp in human beings. Single-stranded DNA occurs as a genetic material in some viruses (e.g., phage ϕ x 174, coliphage fd, M_{13}). DNA is the largest macromolecule with a diameter of 2 nm (20Å or 2×10^{-9}m) and often having 3 length in millimetres.

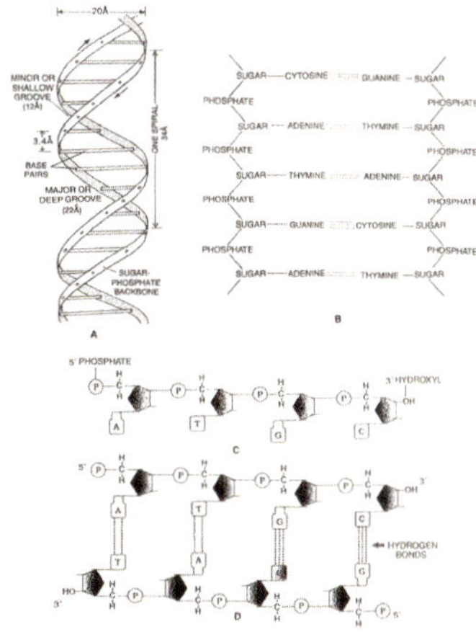

Figure: Structure of DNA. A, coilling in double helix or duplex of DNA. B, sequence of nucleotides in a part of the double helix of DNA. c, a single stranded polynucleotide chain. D, a double stranded polynucleotide chain.

It 15 negatively charged due to phosphate groups. It is a long chain polymer of generally several hundred thousands of deoxyribonucleotides. A DNA molecule has two un-branched complementary strands. They are spirally coiled. The two spiral strands of DNA are collectively called DNA duplex (figure above).

The two strands are not coiled upon each other but the whole double strand (DNA duplex) is coiled upon itself around a common axis like a rope stair case with solid steps twisted into a spiral. Due to spiral twisting, the DNA duplex comes to have two types of alternate grooves, major (22 Å) and minor (12 Å).

In B-DNA, one turn of the spiral has about 10 nucleotides on each strand of DNA. It occupies a distance of about 3.4 nm (34 Å or 3.4×10^{-9} m) so that adjacent nucleotides or their bases are separated by a space of about 0.34 nm (0.34×10^{-9} m or 3.4 Å).

A deoxyribonucleotide of DNA is formed by cross-linking of three chemicals ortho- phosphoric acid (H_3PO_4), deoxyribose sugar ($C_5H_{10}O_4$) and a nitrogen base. Four types of nitrogen bases occur in DNA. They belong to two groups, purines (9-membered double rings with nitrogen at 1,3,7 and 9 positions) and pyrimidines (six membered rings with nitrogen at 1 and 3 positions). DNA has two types of purines (adenine or A and guanine or G) and two types of pyrimidines (cytosine or C and thymine or T).

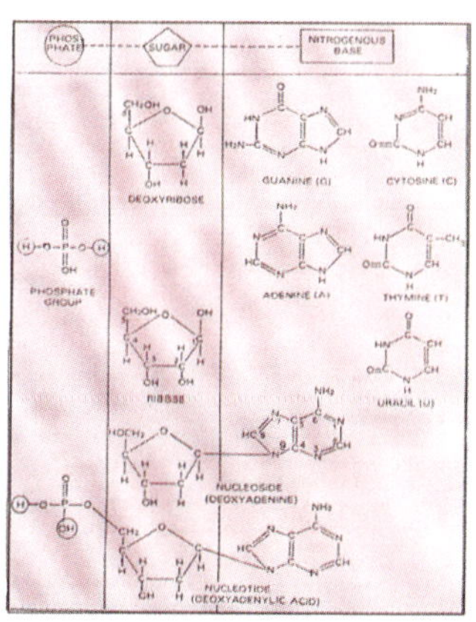

Figure: Building blocks of nuclecotides

Depending upon the type of nitrogen base, DNA has four kinds of deoxyribonucleotides —deoxy adenosine 5- monophosphate (d AMP), deoxy guaninosine 5-monophosphate (d GMP), deoxy thymidine 5-monophosphate (d TMP) and deoxy cytidine 5- monophosphate (d CMP).

The back bone of a DNA chain or strand is built up of alternate deoxyribose sugar and phosphoric acid groups. The phosphate group is connected to carbon 5′ of the sugar residue of its own nucleotide and carbon Y of the sugar residue of the next nucleotide by (3 '—5') phosphodiester bonds. -H of phosphate and -OH of sugar are eliminated as H_2o during each ester formation.

Phosphate group provides acidity to the nucleic acids because at least one of its side group is free to dissociate. Nitrogen bases lie at right angles to the longitudinal axis of DNA chains. They are attached to carbon atom 1 of the sugars by N-glycosidic bonds. Pyrimidine (C or T) is attached to deoxyribose by its N-atom at 1 position while a purine (A or G) does so by N-atom at 9 position.

The two DNA chains are antiparallel that is, they run parallel but in opposite directions. In one chain the direction is 5'→ Y while in the opposite one it is 3′ →5′. The two chains are held together by hydrogen bonds between their bases. Adenine (A), a purine of one chain lies exactly opposite thymine (T), a pyramidine of the other chain. Similarly, cytosine (C, a pyrimidine) lies opposite guanine (G a purine). This allows a sort of lock and key arrangement between large sized purine and small sized pyrimidine.

It is strengthened by the appearance of hydrogen bonds between the two. Three hydrogen bonds occur between cytosine and guanine (C = G) at positions 1'-1', 2′- 6′ and 6′-2′. There are two such hydrogen bonds between adenine and thymine (A=T) which are formed at positions 1'-3′ and 6′-4′. Hydrogen bonds occur between hydrogen of one base and oxygen or nitrogen of the other base. Since specific and different nitrogen bases occur on the two DNA chains, the latter are complementary.

Thus the sequence of say AAGCTCAG of one chain would have a complementary sequence of TTC-GAGTC on the other chain. In other words, the two DNA chains are not identical but complementary to each other. It is because of specific base pairing with a purine lying opposite a pyrimidine. This makes the two chains 2 nm thick.

A purine- purine base pair will make it thicker while a pyrimidine- pyrimidine base pair will make it narrower than 2 nm. Further, A and C or G and T do not pair because they fail to form hydrogen bonds between them. 5′ end of each chain bears phosphate radical while the 3′ end possesses a sugar residue (3'-OH).

Salient Features of B model of DNA of Watson and Crick:

1. DNA is the largest biomolecule in the cell.
2. DNA is negatively charged and dextrorotatory.
3. Molecular configuration of DNA is 3D.
4. DNA has two polynucleotide chains.
5. The two chains of DNA have antiparallel polarity, 5′ —> 3′ in one and 3′ —> 5′ in other.
6. Backbone of each polynucleotide chain is made of alternate sugar-phosphate groups. The nitrogen bases project inwardly.

7. Nitrogen bases of two polynucleotide chains form complementary pairs, A opposite T and C opposite G.

8. A large sized purine always comes opposite a small sized pyrimidine. This generates uniform distance between two strands of helix.

9. Adenine (A) of one polynucleotide chain is held to thymine (T) of opposite chain by two hydrogen bonds. Cytosine (C) of one chain is similarly held to guanine of the other chain by three hydrogen bonds.

10. The double chain is coiled in a helical fashion. The coiling is right handed. This coiling produces minor and major grooves alternately.

11. The pitch of helix is 3.4 nm (34 A) with roughly 10 base pairs in each turn. The average distance between two adjacent base pairs comes to about 0.34 nm (0.34×10^{-9} m or 3.4 A).

12. Planes of adjacent base pairs are stacked over one another. Alongwith hydrogen bonding, the stacking confers stability to the helical structure.

13. DNA is acidic. For its compaction, it requires basic (histone) proteins. The histone proteins are +vely charged and occupy the major grooves of DNA at an angle of 30° to helix axis.

- Sense and Antisense Strands

Both the strands of DNA do not take part in controlling heredity and metabolism. Only one of them does so. The DNA strand which functions as template for RNA synthesis is known as template strand, minus (-) strand or antisense strand.

Its complementary strand is named nontemplate strand, plus (+) strand, sense and coding strand. The latter name is given because by convention DNA genetic code is written according to its sequence.

(5’) GCATTCGGCTAGTAAC (3′)

DNA Nontemplate, Sense (+) or Coding Strand

(3′) CGTAAGCCGATCATTG (5′)

DNA Template, Antisense, or Noncoding or (-) Strand

(5′) GCAUUCGGCUAGUAAC (3′)

RNA Transcript

RNA is transcribed on 3’→5′ (-) strand (template/antistrand) of DNA in 5 → 3 direction.

The term antisense is also used in wider prospective for any sequence or strand of DNA (or RNA) which is complementary to mRNA.

- Denaturation and Renaturation

The H-bonds between nitrogen bases of two strands of DNA can break due to high temperature (82-90°C) or low or high pH, so that the two strands separate from each other. It is called denaturation or melting. Since A-T base pair has only 2H bonds, the area rich in A-T base pairs can undergo easy denaturation (melting). These areas are called low melting areas because they denature at comparatively low temperature. The area rich in G- C base pairs (called high melting area) is comparatively more stable and dense because three hydrogen bonds connect the G-C bases.

These areas have high temperature of melting (Tm). On melting the viscosity of DNA decreases. The denatured DNA has the tendency to reassociate, i.e., the DNA strands separated by melting at 82-90°C can reassociate and form duplex on cooling to temperature at 65°C. It is called renaturation or annealing.

Denatured or separated DNA strands absorb more light energy than the intact DNA double strand. The increased absorption of light energy by separated or denatured DNA strands is called hyperchromatic effect. The effect is used in knowing whether DNA is single or double stranded.

- Palindromic DNA

DNA duplex possesses areas where sequence of nucleotides is the same but opposite in the two strands. These sequences are recognised by restriction endonucleases and are used in genetic engineering. Given hereunder sequence of bases in one strand ($3' \rightarrow 5'$) is GAATTC. It is same in other strand when read in $5' \rightarrow 3'$ direction. It is similar to palindrome words having same words in both forward and backward direction, e.g., NITIN, MALAYALAM.

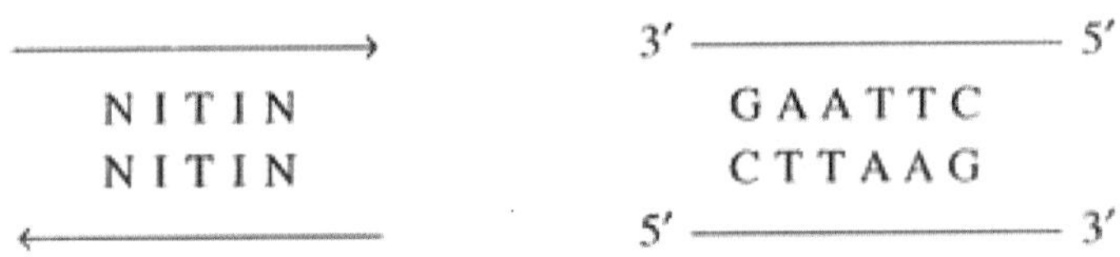

- Repetitive DNA

It is the DNA having multiple copies of identical sequences of nitrogen bases. The number of copies of the same base sequence varies from a few to millions. DNA having single copy of base sequences is called unique DNA. It is made of functional genes. rRNA genes are, however, repeated several times. Repetitive DNA may occur in tandem or interspersed with unique sequences.

It is of two types, highly repetitive and moderately repetitive. Highly repetitive DNA consists of short sequences of less than 10 base pairs which are repeated millions of times. They occur in preeentromeric regions, heterochromatic regions of Y-chromosomes and satellite regions. Moderately repetitive dna consists of a few hundred base pairs repeated at least 1000 times. It occurs in telomeres, centromeres and transposons.

Tandemly repeated sequences are especially liable to undergo misalignments during chromosome pairing, and thus the size of tandem clusters tends to be highly polymorphic, with wide variations between individuals. Smaller clusters of such sequences can be used to characterize individual genomes in the technique of "DNA-finger-printing".

- Satellite DNA

It is that part of repetitive DNA which has long repetitive nucleotide sequences in tandem that forms a separate fraction on density ultracentrifugation. Depending upon the number of base pairs involved in repeat regions, satellite DNA is of two types, microsatellite sequences (1-6 bp repeat units flanked by conserved sequences) and minisatellite sequences (11-60 bp flanked by conserved restriction sites). The latter are hyper variable and are specific for each individual. They are being used for DNA matching or finger printing as first found out by Jeffreys.

- Genetic Information

The arrangement of nitrogen bases of DNA (and its product mRNA) determines the sequence of amino acid groups in polypeptides or proteins formed over ribosomes. One amino acid is specified by the sequence of three adjacent nitrogen bases. The latter is called codon. The segment of DNA which determines the synthesis of complete polypeptide is known as cistron.

In procaryotes, a cistron has a continuous coding sequence from beginning to end. In eucaryotes a cistron contains noncoding regions which do not produce part of gene product. They are called introns. Introns are often variable. The coding parts are known as exons. Cistrons having introns are called split genes.

- Coding and Noncoding DNA

Depending on the ability to form functional or nonfunctional products, DNA has two types of segments, coding and noncoding. In eukaryotes a greater part of DNA is noncoding since it does not form any functional product. They often possess repeated sequences or repetitive DNA. Most of them have fixed positions.

Some can move from one place to another. The mobile sequences are called jumping genes or transposons. In prokaryotes the amount of noncoding or nonfunctional DNA is small. Coding DNA consists of coding DNA sequences. These are of 2 types — protein coding sequences coding for all proteins except histone and nonprotein coding sequences for tRNA, rRNA and histones.

Functions of DNA

1. Genetic Information (Genetic Blue Print):

DNA is the genetic material which carries all the hereditary information. The genetic information is coded in the arrangement of its nitrogen bases.

2. Replication:

DNA has unique property of replication or production of carbon copies (Autocatalytic function). This is essential for transfer of genetic information from one cell to its daughters and from one generation to the next.

3. Chromosomes:

DNA occurs inside chromosomes. This is essential for equitable distribution of DNA during cell division.

4. Recombination's:

During meiosis, crossing over gives rise to new combination of genes called recombinations.

5. Mutations:

Changes in sequence of nitrogen bases due to addition, deletion or wrong replication give rise to mutations. Mutations are the fountain head of all variations and evolution.

6. Transcription:

DNA gives rise to RNAs through the process of transcription. It is heterocatalytic activity of DNA.

7. Cellular Metabolism:

It controls the metabolic reactions of the cells through the help of specific RNAs, synthesis of specific proteins, enzymes and hormones.

8. Differentiation:

Due to differential functioning of some specific regions of DNA or genes, different parts of the organisms get differentiated in shape, size and functions.

9. Development:

DNA controls development of an organism through working of an internal genetic clock with or without the help of extrinsic information.

10. DNA Finger Printing:

Hypervariable microsatellite DNA sequences of each individual are distinct. They are used in identification of individuals and deciphering their relationships. The mechanism is called DNA finger printing.

11. Gene Therapy:

Defective heredity can be rectified by incorporating correct genes in place of defective ones.

12. Antisense Therapy:

Excess availability of anti-mRNA or antisense RNAs will not allow the pathogenic genes to express themselves. By this technique failure of angioplasty has been checked. A modification of this technique is RNA interference (RNAi).

RNA

RNA is abbreviation of ribonucleic acid which is a complex compound of high molecular weight that functions in cellular protein synthesis and replaces DNA (deoxyribonucleic acid) as a carrier of genetic codes in some viruses.

RNA consists of ribose nucleotides (nitrogenous bases appended to a ribose sugar) attached by phosphodiester bonds, forming strands of varying lengths. The nitrogenous bases in RNA are adenine, guanine, cytosine, and uracil, which replaces thymine in DNA.

The ribose sugar of RNA is a cyclical structure consisting of five carbons and one oxygen. The presence of a chemically reactive hydroxyl (–OH) group attached to the second carbon group in the ribose sugar molecule makes RNA prone to hydrolysis. This chemical lability of RNA, compared with DNA, which does not have a reactive –OH group in the same position on the sugar moiety (deoxyribose), is thought to be one reason why DNA evolved to be the preferred carrier of genetic information in most organisms. The structure of the RNA molecule was described by R.W. Holley in 1965.

RNA Structure

RNA typically is a single-stranded biopolymer. However, the presence of self-complementary sequences in the RNA strand leads to intrachain base-pairing and folding of the ribonucleotide chain into complex structural forms consisting of bulges and helices. The three-dimensional structure of RNA is critical to its stability and function, allowing the ribose sugar and the nitrogenous bases to be modified in numerous different ways by cellular enzymes that attach chemical groups (e.g., methyl groups) to the chain. Such modifications enable the formation of chemical bonds between distant regions in the RNA strand, leading to complex contortions in the RNA chain, which further stabilizes the RNA structure. Molecules with weak structural modifications and stabilization may be readily destroyed. As an example, in an initiator transfer RNA (tRNA) molecule that lacks a methyl group ($tRNA_i^{Met}$), modification at position 58 of the tRNA chain renders the molecule unstable and hence nonfunctional; the non-functional chain is destroyed by cellular tRNA quality control mechanisms.

RNAs can also form complexes with molecules known as ribonucleoproteins (RNPs). The RNA portion of at least one cellular RNP has been shown to act as a biological catalyst, a function previously ascribed only to proteins.

Types And Functions of RNA

Of the many types of RNA, the three most well-known and most commonly studied are messenger RNA (mRNA), transfer RNA (tRNA), and ribosomal RNA (rRNA), which are present in all organisms. These and other types of RNAs primarily carry out biochemical reactions, similar to enzymes. Some, however, also have complex regulatory functions in cells. Owing to their involvement in many regulatory processes, to their abundance, and to their diverse functions, RNAs play important roles in both normal cellular processes and diseases.

In protein synthesis, mRNA carries genetic codes from the DNA in the nucleus to ribosomes, the sites of protein translation in the cytoplasm. Ribosomes are composed of rRNA and protein. The ribosome protein subunits are encoded by rRNA and are synthesized in the nucleolus. Once fully assembled, they move to the cytoplasm, where, as key regulators of translation, they "read" the code carried by mRNA. A sequence of three nitrogenous bases in mRNA specifies incorporation of a specific amino acid in the sequence that makes up the protein. Molecules of tRNA (sometimes also called soluble, or activator, RNA), which contain fewer than 100 nucleotides, bring the specified amino acids to the ribosomes, where they are linked to form proteins.

In addition to mRNA, tRNA, and rRNA, RNAs can be broadly divided into coding (cRNA) and noncoding RNA (ncRNA). There are two types of ncRNAs, housekeeping ncRNAs (tRNA and rRNA) and regulatory ncRNAs, which are further classified according to their size. Long ncRNAs (lncRNA) have at least 200 nucleotides, while small ncRNAs have fewer than 200 nucleotides. Small ncRNAs are subdivided into micro RNA (miRNA), small nucleolar RNA (snoRNA), small nuclear RNA (snRNA), small-interfering RNA (siRNA), and PIWI-interacting RNA (piRNA).

The miRNAs are of particular importance. They are about 22 nucleotides long and function in gene regulation in most eukaryotes. They can inhibit (silence) gene expression by binding to target mRNA and inhibiting translation, thereby preventing functional proteins from being produced. Many miRNAs play significant roles in cancer and other diseases. For example, tumour suppressor and oncogenic (cancer-initiating) miRNAs can regulate unique target genes, leading to tumorigenesis and tumour progression.

Also of functional significance are the piRNAs, which are about 26 to 31 nucleotides long and exist in most animals. They regulate the expression of transposons (jumping genes) by keeping the genes from being transcribed in the germ cells (sperm and eggs). Most piRNA are complementary to different transposons and can specifically target those transposons.

Circular RNA (circRNA) is unique from other RNA types because its 5′ and 3′ ends are bonded together, creating a loop. The circRNAs are generated from many protein-encoding genes, and some can serve as templates for protein synthesis, similar to mRNA. They can also bind miRNA, acting as "sponges" that prevent miRNA molecules from binding to their targets. In addition, circRNAs play an important role in regulating the transcription and alternative splicing of the genes from which circRNAs were derived.

RNA in Disease

Important connections have been discovered between RNA and human disease. For example, as described previously, some miRNAs are capable of regulating cancer-associated genes in ways that facilitate tumour development. In addition, the dysregulation of miRNA metabolism has been linked to various neurodegenerative diseases, including Alzheimer disease. In the case of other RNA types, tRNAs can bind to specialized proteins known as caspases, which are involved in apoptosis (programmed cell death). By binding to caspase proteins, tRNAs inhibit apoptosis; the ability of cells to escape programmed death signaling is a hallmark of cancer. Noncoding RNAs known as tRNA-derived fragments (tRFs) are also suspected to play a role in cancer. The emergence of techniques such as RNA sequencing has led to the identification of novel classes of tumour-specific RNA transcripts, such as MALAT1 (metastasis associated lung adenocarcinoma transcript 1), increased levels of which have been found in various cancerous tissues and are associated with the proliferation and metastasis (spread) of tumour cells.

A class of RNAs containing repeat sequences is known to sequester RNA-binding proteins (RBPs), resulting in the formation of foci or aggregates in neural tissues. These aggregates play a role in the development of neurological diseases such as amyotrophic lateral sclerosis (ALS) and myotonic dystrophy. The loss of function, dysregulation, and mutation of various RBPs has been implicated in a host of human diseases.

The discovery of additional links between RNA and disease is expected. Increased understanding of RNA and its functions, combined with the continued development of sequencing technologies and efforts to screen RNA and RBPs as therapeutic targets, are likely to facilitate such discoveries.

Chromosomes

A chromosome is a string of DNA wrapped around associated proteins that give the connected nucleic acid bases a structure. During interphase of the cell cycle, the chromosome exists in a loose structure, so proteins can be translated from the DNA and the DNA can be replicated. During mitosis and meiosis, the chromosome becomes condensed, to be organized and separated. The substance consisting of all the chromosomes in a cell and all their associated proteins is known as chromatin. In prokaryotes, there is usually only a single chromosome, which exists in a ring-like or linear shape. The chromatin of most eukaryotic organisms consists of multiple chromosomes. Each chromosome carries part of the genetic code necessary to produce an organism.

Having the entire genetic code divided into different chromosomes allows the possibility of variation through the different combinations of chromosomes with the different alleles, or genetic variations that they contain. The recombination and mutation of chromosomes can occur during mitosis, meiosis, or during interphase. The end result is organisms that function and behave in different ways. This variation allows populations to evolve over time, in response to changing environments.

Function of a Chromosome

The chromosome holds not only the genetic code, but many of the proteins responsible for helping express it. Its complex form and structure dictate how often genes can be translated into proteins, and which genes are translated. This process is known as gene expression and is responsible for creating organisms. Depending on how densely packed the chromosome is at certain point determines how often a gene gets expressed. As seen in the image of chromosome structure shown below, less active genes will be more tightly packed than genes undergoing active transcription. Cellular molecules that regulate genes and transcription often work by activing or deactivating these proteins, which can contract or expand the chromosome. During cell division, all the proteins are activated and the chromatin becomes densely packed into distinct chromosomes. These dense molecules have a better chance of withstanding the pulling forces that occur when chromosomes are separated into new cells.

Chromosome Structure

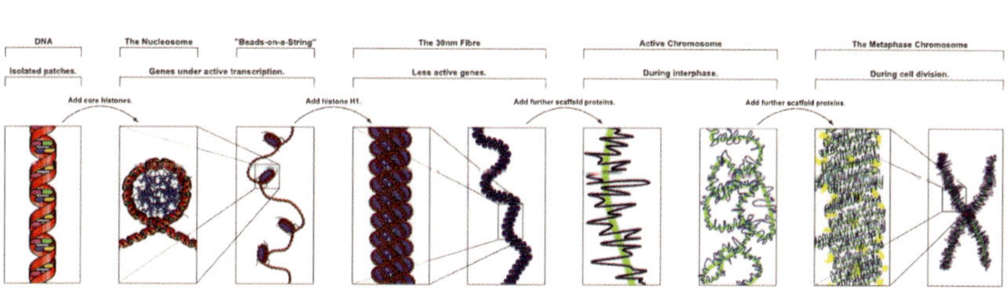

Chromosomes have a very complex structure. DNA, or deoxyribonucleic acid makes the base of the structure, as seen on the far left. DNA is made of a two strings of nucleic acid base pairs. The base pairs in DNA are cytosine, adenine, thymine, and guanine. The spiral structure formed by the two strings of DNA is due to complimentary pairing between every base with its pair on the opposite string. Adenine pairs with thymine and guanine pairs with cytosine. The opposite side of the bases form a phosphate-deoxyribose backbone, which keeps the strands intact.

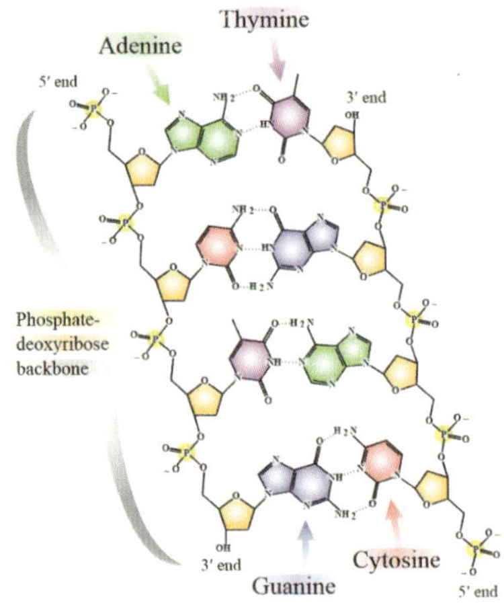

DNA chemical structure

When the DNA is duplicated, the strands are separated, and a polymerase molecule builds a new string that corresponds to each side. In this way, the DNA is perfectly replicated. This can be done artificially by a polymerase chain reaction in which special enzymes and heat are used to separate and replicated the strings a number of times, to produce many copies of the same DNA. This makes it much easier to study any string of DNA, even whole chromosomes or genomes.

After the cell has expressed and duplicated the DNA, cell division can occur. This occurs in both prokaryotes and eukaryotes, but only eukaryotes condense their DNA so it can be separated. Prokaryotic DNA is so simple that relatively few structural proteins are associated with the chromosome. In eukaryotes, many structural proteins are used.

The first of these proteins are core histones. Many individual histone proteins bind together to form a core histone. The DNA can wrap around one of these histones, giving it a wound structure. This structure, and the associated histone, is known as the nucleosome. As seen in the third picture from the left, these nucleosomes form "beads-on-a-string". The string becomes wound back and forth by another histone, histone H1, and eventually fibers are produced. The next type of protein, scaffold proteins, start to wind the fiber into a loose structure. When the chromosome must condense during cell division, more scaffold proteins are activated, and the structure becomes much denser. In fact, even with a microscope, individual chromosomes cannot be discerned until near the middle of cell division cycles, when the chromosome becomes very dense. This process is seen as the pictures progress towards the right.

Bacterial Chromosome

Bacterial chromosomes contain circular DNA molecule unlike the linear DNA of vertebrates. Most of chromosomes are circular DNA molecules and there are no free ends to the DNA. The bacterial DNA is packaged into a single chromosome into a continuous loop. The DNA is folded or coiled to fit into the cell. The compaction of the DNA involves the binding of proteins to the DNA that help form initial loops which is then coiled.

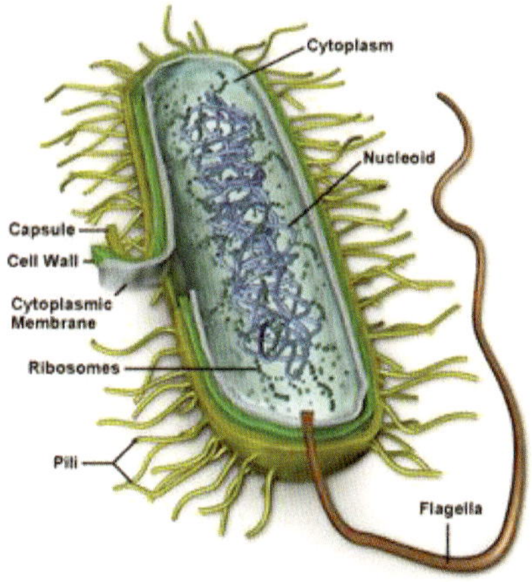

Prokaryotic Chromosome

Prokaryotes like the bacteria and archaea typically have a single circular chromosome. The chromosome size of most bacteria is from only 160,000 base pairs to 12,200,00 base pairs.Some bacteria in exceptions contain a single linear chromosome. The base sequences in prokaryotic chromosomes are less than in eukaryotic cells. Bacterial chromosomes have a single origin of replication from which the replication starts. In some archaea there are multiple replication origins. The prokaryotic genes are organized into operons and it usually it does not contain introns. Nucleus is absent in prokaryotes, the DNA is organized into a structure called the nucleoid. The DNA of the archaea are more organized, they are packaged within structures similar to eukaryotic nucleosomes. The chromosomes in the prokaryotes and plasmids are generally supercoiled like that of the eukaryotes. The DNA are released into the relaxed state for the process of transcription, replication and regulation.

Eukaryotic Chromosomes

In eukaryotes the chromosomes are multiple large, linear and are present in the nucleus of the cell. Each chromosome typically has one centromere, one or two arms that project from the centromere, the arms are usually not visible during most of the time. Most of the eukaryotes have a small circular genome in the mitochondria. Some of the eukaryotes have small linear or circular chromosomes in the cytoplasm. To fit into the compartment in which it is contained the DNA has to be condensed and the degree to which it is condensed is expressed as its packaging ratio. Packaging ratio is the length of the DNA divided by the length into which it is packaged. The chromosomes of eukaryotic organisms consists of complexes made of DNA and protein, and it is organized in a condensed manner. This condensation permits the large amount of DNA to be stored in the nucleus of the cell.

Human Chromosomes

Humans chromosomes are of two types autosomes and sex chromosomes. Genetic traits that are linked to the sex of the person are passed on through the sex chromosomes. The rest of the genetic information is present in the autosomes. Humans have 23 pairs of chromosomes in their cells, of which 22 pairs are autosomes and one pair of sex chromosomes, making a total of 46 chromosomes in each cell. Many copies of mitochondrial genome are present in human cells.

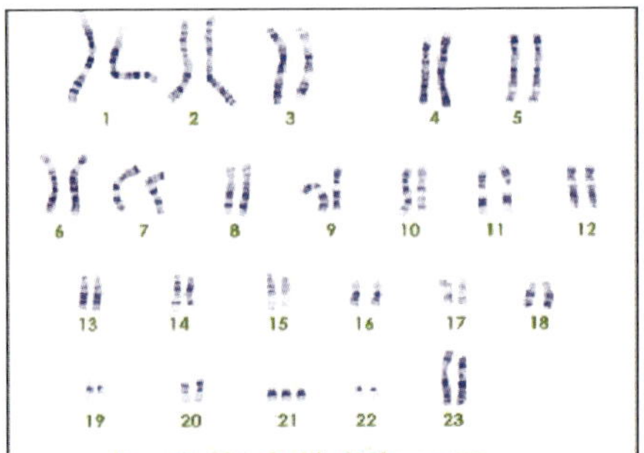

Human Karyotype Contains 46 Chromosomes

Sex Chromosomes

Sex chromosomes differ in form of size, behavior from the ordinary chromosome. The sex chromosomes determine the sex of an individual during reproduction. These sex chromosomes differ between the male and the females. Females have two copies of X chromosome, males have one X chromosome and one Y chromosome. In the process of sexual reproduction in humans, two different gametes fuse to form a zygote.

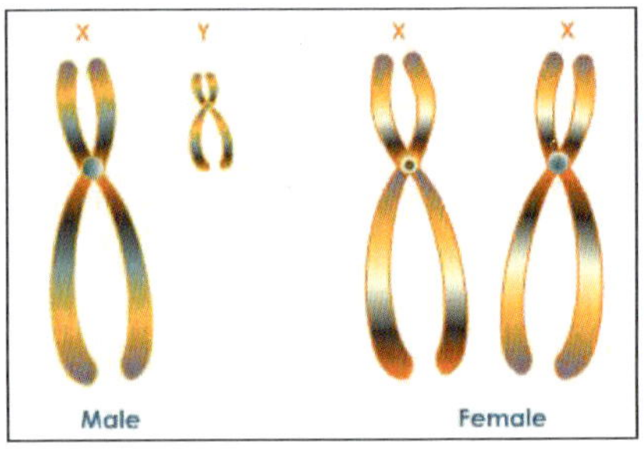

Homologous Chromosomes

Homologous chromosomes are also known as homologs or homologues. The homologous chromosomes are pairs of chromosomes that are approximately of the same length, position of centromere, and pattern of staining, genes for the same characteristic are at a corresponding loci. In an organism one of the homologous chromosome is inherited from the mother and the other from the father. These chromosomes are usually not identical, but they carry the same type of genes. During the process of mitosis the daughter chromosomes carry the same sequence of nucleotide, assuming there are no errors during the replication process. The genome in diploid organisms is composed of homologous chromosomes One of homologous pair is the maternal chromosome and the other is the paternal chromosome. During the process of meiosis the homologous chromosomes cross over. Homologous chromosomes are not identical but they are similar. The genes are carried in the same order, but the alleles for the trait may not be similar.

Meiosis

Meiosis is a specialized form of cell division process in which a single cell divided twice to form four daughter cells that possess exactly half the number of chromosomes of the parent cell. Therefore, all the four cells are haploid and will contain half the original amount of the genetic information. These cells are call sex cells: sperms in male whereas eggs in females.

Stages of Meiosis

Meiosis can be majorly categorized into 9 different stages. As there are two divisions in meiosis, the two divisions are called meiosis I (where the cell divides for the first time) and meiosis II (where the cell divides for the second time).

Meiosis starts with duplication of chromosomes. Then the parent cell undergoes two rounds of nuclear divisions (meiosis I and meiosis II). Thus in an overall meiosis process, four daughter cells are produced from the parent cell. However, before entering the stages of meiosis the parent cell go through a phase called interphase. Similar to mitosis, the cells growth begins in the G1 phase of interphase, the chromosomes are duplicated in the S phase and then in G2 phase the parent cells prepares for cell division.

Before the start of meiosis I, the chromosomes are duplicated and they are fused together. The duplicated chromosomes are called sister chromatids. The point at which the duplicated chromosomes (sister chromatids) are joined is called centromere. The complete structure looks like a "letter X". The chromosomes condense and become compact at the time of every nuclear division. At this point, they are visible under the microscope.

Meiosis I

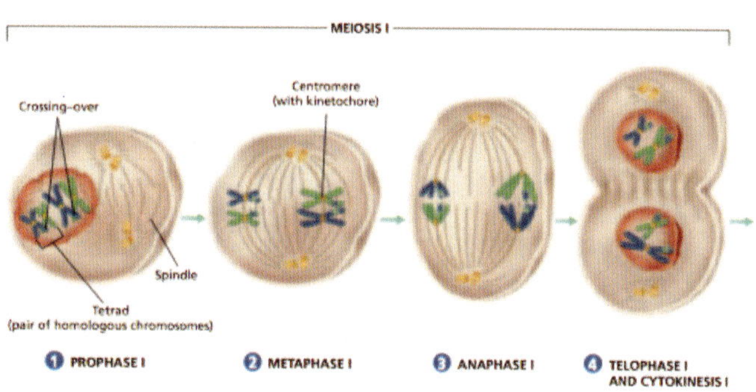

In this stage, the sister chromatids or the chromosomes of the maternal set combines together with their homologs of the paternal chromosomes set. The duplicated chromosomes condense and in combination, they look like two X's present next to each other.

Every chromosome is made up of two sister chromatids that contain similar genetic information. The chromosomes pair up in a pattern that copies of particular chromosomes are paired together. The maternal and paternal chromatids exchange their parts of DNA and recombine to develop genetic variations.

For instance, in the image below, the letters A, B, and C represent genes found at particular spots on the chromosome, with capital and lowercase letters for different forms, or alleles, of each gene. The DNA is broken at the same spot on each homologue—here, between genes B and C—and reconnected in a criss-cross pattern so that the homologues exchange part of their DNA.

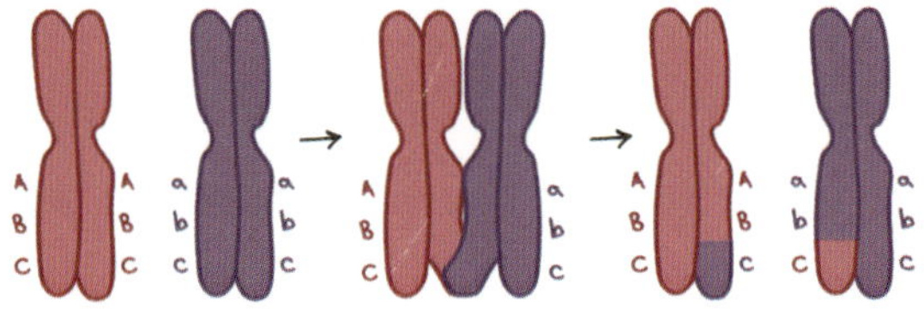

The process of exchange of genetic information is called recombination or crossing over. However, in the case of a male in humans the sex chromosomes X and Y are not homologs but they can pair together and exchange DNA. Recombination happens in a small part of two chromosomes where homology is present.

The chromosomes would actually be positioned one on top of the other—as in the image below—throughout crossing over; they're only shown side-by-side in the image above so that it's easier to see the exchange of genetic material.

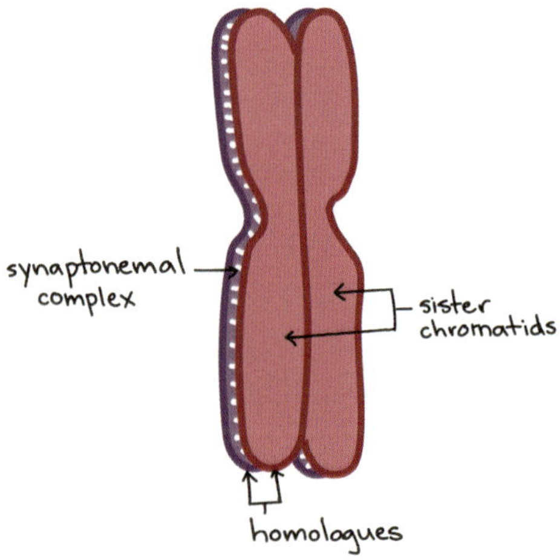

The nuclear membrane of the cell breaks down releasing the chromosomes by the end of prophase I. The meiotic spindle that constitutes microtubules and other proteins that is stretched across the cell in between the centrioles.

Metaphase I

The chromosome sets arrange themselves next to each other along the equator of the cell. The meiotic spindle is positioned on both the sides of the cell. At that time the centrioles are present in the opposite ends of the cell and the meiotic spindles stretch from them.

The meiotic spindle attaches to the combined sister chromatids (one chromosome of each pair). At the end of metaphase I, the fused sister chromatids are latched to the centromere and line up in the center of the cell. The homologs still represent X sitting next to each other.

Anaphase I

The chromosome pairs are gradually pulled apart when the meiotic spindle fibres start to contract. As a result, one chromosome is pulled to one end of the cell and the other chromosome to the opposite end. Thus each X shape structure moves away from each other towards the opposite pole.

Telophase I and Cytokinesis

In this stage, the chromosomes completely move apart and are present in the opposite poles of the cell. Each pole of the cell contains a full set of chromosomes arrange together. A membrane develops around each chromosome pair. At this point, the single cell splits in the middle to create two separate daughter cells by a process called cytokinesis.

Each daughter cell is composed of a full set of chromosomes present inside the nucleus. At the end of Meiosis I, two daughter cells each having a full set of chromosomes are present. The genetic makeup of each newly formed cells is different because of the DNA exchange taking place between the homologs earlier.

The newly formed daughter cells enter into meiosis II without further duplication of chromosomes.

Meiosis II

Meiosis II process is similar to an equational division. Thus by the end of meiosis II, the chromosome number of cells that are undergoing the meiosis II process and in the resultant daughter cells remain unchanged. There are four stages of meiosis II.

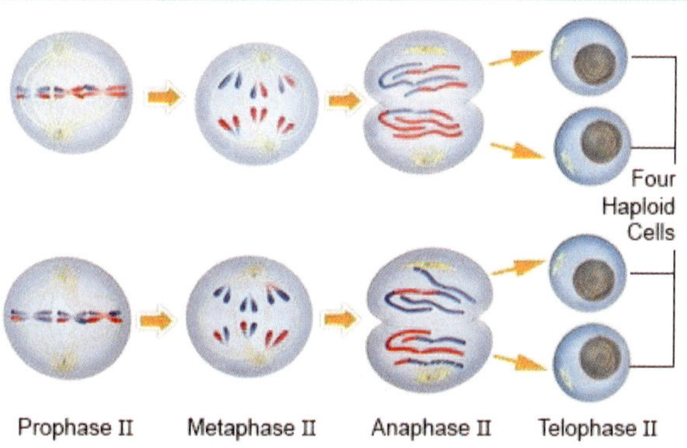

Prophase II

After meiosis I, there are two cells with the same number of chromosomes or the same number of chromatid pairs. The chromosomes in each daughter cell once again condense into visible X shaped structure. The nuclear membrane around the nucleus of the daughter cells disintegrates releasing the chromosome. The meiotic spindles start forming again and the centrioles duplicate.

Metaphase II- The pair of sister chromatids arrange themselves in line along the equator of the cell. The centrioles are now present at the opposite ends in each of the daughter cells. The developed meiotic spindle fibres at each end attach onto the centromere present in the sister chromatids of the cell.

Anaphase II

The meiotic spindle fibres start contracting and the sister chromatids are pulled to the opposite ends. At this point, the chromatids are separated and they form individual chromosomes.

Telophase II

The chromosomes finally move to the opposite ends of the cell. At each end or pole, an entire set of chromosomes are arranged together. Again the nucleus membrane develops around each pair of chromosomes and two new cell nuclei are formed. This marks the last phase of meiosis but the cell division process remains incomplete without the cytokinesis process. So the cell undergoes cytokinesis and after the process, four granddaughter cells are formed.

Each newly formed cell contains half a set of chromosomes or haploid chromosomes. Each chromosome is distinct and unique. It is composed of a mix of genetic material from the paternal as well as paternal chromosomes of the original cell.

Mixing and Matching of Genes During Meiosis

The gametes produced in meiosis are all haploid, but they're not genetically identical. For example, take a look the meiosis II diagram above, which shows the products of meiosis for a cell with 2n = 4 chromosomes. Each gamete has a unique "sample" of the genetic material present in the starting cell.

As it turns out, there are many more potential gamete types than just the four shown in the diagram, even for a cell with only four chromosomes. The two main reasons we can get many genetically different gametes are:

- Crossing over. The points where homologues cross over and exchange genetic material are chosen more or less at random, and they will be different in each cell that goes through meiosis. If meiosis happens many times, as in humans, crossovers will happen at many different points.
- Random orientation of homologue pairs. The random orientation of homologue pairs in metaphase I allows for the production of gametes with many different assortments of homologous chromosomes.

In a human cell, the random orientation of homologue pairs alone allows for over 8 million, n different types of possible gametes When we layer crossing over on top of this, the number of genetically different gametes that you—or any other person—can make is effectively infinite.

Mitosis

Mitosis is the step in the cell cycle that the newly duplicated DNA is separated, and two new cells are formed. This process is important in single-celled eukaryotes, as it is the process of asexual reproduction. In multi-celled eukaryotes, mitosis is how a single zygote can become an entire organism. Mitosis has several distinct stages, or phases, that will be discussed below. The other stages in the cell cycle include growth and the replication of DNA, both required for mitosis to take place. Below is a picture of where mitosis fits into the cell cycle.

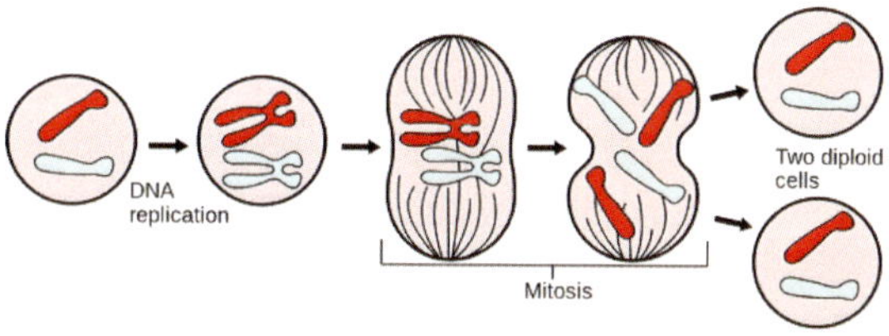

Functions of Mitosis

Development

In multi-celled organisms, life always starts as a single cell, formed from two gametes. This zygote carries all the DNA necessary to create a fully functioning organism, but not near enough cells. The purpose of mitosis is to produce more cells. After the first round of mitosis, there are only two cells. These cells both undergo mitosis, and there are 4 cells. Pretty soon, a small, hollow ball of cells is formed, called the blastula. This ball folds in on itself as more and more cells are created. The cells begin to differentiate, which allows them to complete special tasks in the body. Eventually, a fully functioning organism is developed and can be birthed or hatched into the world.

In single-celled organisms, the act of mitosis is asexual reproduction. Single-celled organisms use mitosis to reproduce and distribute their DNA. Some single-celled organisms reproduce sexually as well. To reproduce sexually, most organisms undergo another process, meiosis, to properly reduce their DNA and place the DNA in individual cells. These gametes can then meet and one will become fertilized. This fertilized gamete contains two sets of the genome, which in most organisms is necessary for proper development. Some organisms only have a single copy of the DNA. These are known as diploid and haploid organisms, respectively.

Replacing Damaged Tissues

The second important function of mitosis is that of repair. When an organism gets injured, its cells are damaged. This can be a physical injury like a cut, or damage from environmental sources like the sun. Either way, the damaged cells need replaced. Nearby cells, not sensing their neighbor cells, turn on the pathways that start the process of mitosis. Eventually the multiplying new cells reach each other, and the damage area is covered with new cells. Some organism are able to regenerate entire limbs this way. Lizards, crabs, and many other animals can lose a tail or claw with no fear, as the limb can be regrown through mitosis.

Phases of Mitosis

Though technically not part of mitosis, Interphase begins and ends mitosis. Interphase is the part of the cell cycle in which the cell grows and duplicates the DNA. After an identical set of DNA is synthesized, the cell enters Mitosis.

Prophase

Prophase is the first stage in mitosis. During prophase, the DNA is condensed. During interphase, when the DNA is replicated, it is in a loose and open form to allow the enzymes to do their work on the DNA and create a new strand. However this chromatin, as it is called, would become tangled and break if the cell tried to move it without condensing it. During prophase, the cell's machinery packages the DNA around special proteins, called histones, which allow it to condense into very tight packages. These tight packages of DNA can now be moved around with ease. During prophase the centrioles appear, which are centers on each side of the cell that organize microtubules. The microtubules will eventually reach out and grab the chromosomes of DNA.

In plants this stage is proceeded by a step that rearranges the cell to put the nucleus in the middle. In most animal cells, the nucleus exists in the center of the cell most of the time. In plants, it is often pushed to one side by large, water-containing vacuoles. This preprophase allows plants to organize their organelles for division.

Prometaphase

In order for the duplicated chromosomes to be split up, the microtubules have to reach them. In prometaphase, the nuclear envelope surrounding the cells falls apart. This membrane was separating the DNA from the cytosol of the cell. When the nuclear envelope dissolves, the microtubules are allowed to extend from the centromeres to the chromosomes. Each chromosome has a special area known as the centromere, and each centromere has a kinetochore. The microtubules are able to attach to these kinetochores, allowing the cell to move the chromosomes around. Microtubules from each side of the cell will attach to each chromosome during prometaphase.

Metaphase

During metaphase, the microtubules start pulling on the chromosomes. Each side pulls with equal force, and the chromosomes end up in the middle of the cell. This area is called the metaphase plate. The cells aligned on the metaphase plate represent two full copies of DNA. Each chromosome lines up next to its sister chromatid, or cloned DNA strand. In this way, when the microtubules pull the chromosomes apart, each cell gets an entire functioning genome. Below is a picture of a cell in metaphase.

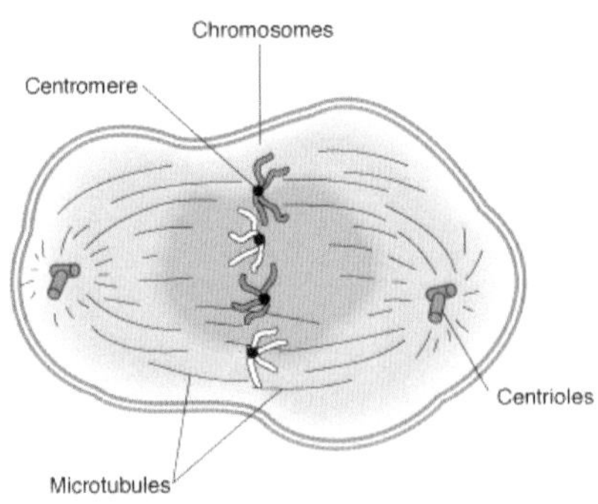

Anaphase

The sister chromatids, identical clones of the same part of DNA, are bound together at their centromeres. During anaphase of mitosis, the proteins that connect these chromatids are destroyed. Each now its own chromosome, the identical halves can be pulled to each cell. In the above picture, there are two white chromosomes and two grey chromosomes. During anaphase the proteins between the two sister chromatids in each chromosome would dissolve. In the above cell this would lead to 8 total chromosomes after the chromatids separate. In the next phase, they will be separated, to create 4 chromosomes in each cell, the number the cell had before it duplicated its DNA.

Telophase

The final phase of Mitosis, telophase, occurs when the chromosomes are pulled toward each centriole, and a cleavage furrow forms in the cell. The chromosomes will eventually get a nuclear envelope formed around them, and become their own cells. The centrioles will dissolve, and each cell will resume normal functioning. One crucial last step, cytokinesis, is needed before the cell function. This last process is not a step of mitosis, but rather the beginning of Interphase. Once divided, the cells can resume growing.

Genetic Inheritance

Biological inheritance is the process by which an offspring cell or organism acquires or becomes predisposed to characteristics of its parent cell or organism. Through inheritance, variations exhibited by individuals can accumulate and cause a species to evolve. The study of biological inheritance is called genetics.

Categories of Biological Inheritance

The description of a mode of biological inheritance consists of three main categories:

1. Number of involved loci:

    i. Monogenetic (also called 'simple')—one locus

    ii. Oligogenetic—few loci

    iii. Polygenetic—many loci

2. Involved chromosomes:

    a) Autosomal—Loci are not situated on a sex chromosome

    b) Gonosomal—Loci are situated on a sex chromosome

    c) X-chromosomal—Loci are situated on the X chromosome (the more common case)

    d) Y-chromosomal—Loci are situated on the Y chromosome

    e) Mitochondrial—Loci are situated on the mitochondrial DNA

3. Correlation genotype-phenotype:

    i. Dominant

    ii. Intermediate (also called 'co-dominant')

    iii. Recessive

These three categories are part of every exact description of a mode of inheritance in the above order.

Specifications of Biological Inheritance

Additionally, more specifications may be added as follows:

1. Coincidental and environmental interactions:
    - i. Penetrance
    - ii. Incomplete (percentual number)
    - iii. Invariable
    - iv. Complete
    - v. Expressivity
    - vi. Variable
    - vii. Maternal or paternal imprinting phenomena
    - viii. Heritability (in polygenetic and sometimes also in oligogenetic modes of inheritance)
2. Sex-linked interactions:
    - A. Sex-linked inheritance (gonosomal loci)
    - B. Sex-limited phenotype expression (ex. Cryptorchism)
    - C. Inheritance through the maternal line (in case of mitochondrial DNA loci)
    - D. Inheritance through the paternal line (in case of Y-chromosomal loci)
3. Locus-locus-interactions:
    - a. Epistasis with other loci (ex. Over-dominance)
    - b. Gene coupling with other loci
    - c. Homozygotes lethal factors
    - d. Semi-lethal factors

Determination and description of a mode of inheritance is primarily achieved through statistical analysis of pedigree data. In case the involved loci are known, methods of molecular genetics can also be employed. Mendelian inheritance (or Mendelian genetics or Mendelism) is a set of primary tenets relating to the transmission of hereditary characteristics from parent organisms to their children.

Mendel's Laws

Law of Segregation

The "Law of Segregation", also known as Mendel's first Law essentially has three parts:

1. Alternative versions of genes account for variations in inherited characteristics:

This is the concept of alleles. Alleles are different versions of genes that impart the same characteristic. For example, each human has a gene that controls eye color, but there are variations among these genes in accordance with the specific color for which the gene 'codes'.

2. For each characteristic, an organism inherits two alleles, one from each parent:

This means that when somatic cells are produced from two alleles, one allele comes from the mother and one from the father. These alleles may be the same (true-breeding organisms/homozygous, ex. W and 'rr' in the figure below), or different (hybrids/heterozygous, ex. 'wr' in figure below).

3. The two alleles for each characteristic segregate during gamete production:

This means that each gamete will contain only one allele for each gene. This allows the maternal and paternal alleles to be combined in the offspring, ensuring variation. It is often misconstrued that the gene itself is dominant, recessive, co-dominant, or incompletely dominant. It is however, the trait or gene product that the allele encodes that is dominant, etc.

Law of Independent Assortment

The law of independent assortment, also known as "Law of Inheritance" or Mendel's Second Law, states that the inheritance pattern of one trait will not affect the inheritance pattern of another. While his experiments with mixing one trait always resulted in a 3:1 ratio between dominant and recessive phenotypes, his experiments with mixing two traits (di-hybrid cross) showed 9:3:3:1 ratios.

But the 9:3:3:1 table shows that each of the two genes are independently inherited with a 3:1 ratio. Mendel concluded that different traits are inherited independently of each other, so that there is no relation, for example, between a cat's color and tail length. This is actually only true for genes that are not linked to each other.

Independent assortment occurs during meiosis-I in eukaryotic organisms, specifically anaphase-I of meiosis, to produce a gamete with a mixture of the organism's maternal and paternal chromosomes. Along with chromosomal cross-over, this process aids in increasing genetic diversity by producing novel genetic combinations.

Of the 46 chromosomes in a normal diploid human cell, half are maternally derived (from the mother's egg) and half are paternally derived (from the father's sperm). This occurs as sexual reproduction involves the fusion of two haploid gametes (the egg and sperm) to produce a new organism having the full complement of chromosomes.

During gametogenesis, the production of new gametes by an adult the normal complement of 46 chromosomes needs to be halved to 23 to ensure that the resulting haploid gamete can join with another gamete to produce a diploid organism. An error in the number of chromosomes, such as those caused by a diploid gamete joining with a haploid gamete, is termed aneuploidy.

In independent assortment the chromosomes that end up in a newly-formed gamete are randomly sorted from all possible combinations of maternal and paternal chromosomes. Because gametes end up with a random mix instead of a pre-defined "set" from either parent, gametes are therefore considered assorted independently. As such, the gamete can end up with any combination of paternal or maternal chromosomes.

Any of the possible combinations of gametes formed from maternal and paternal chromosomes will occur with equal frequency. For human gametes, with 23 pairs of chromosomes, the number of possibilities is 2:23 or 8,388,608 possible combinations. The gametes will normally end up with 23 chromosomes, but the origin of any particular one will be randomly selected from paternal or maternal chromosomes. This contributes to the genetic variability of progeny.

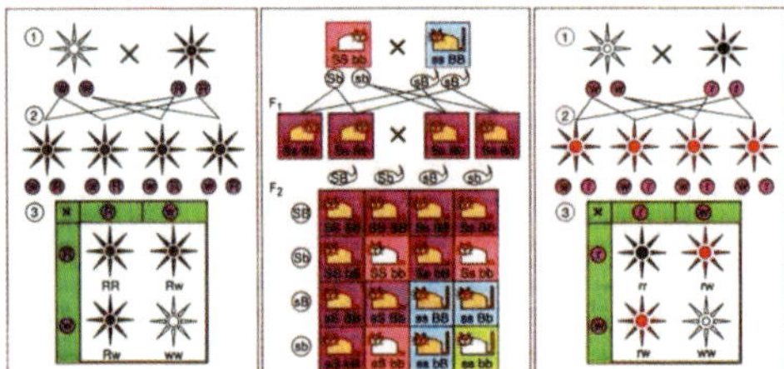

Figure1: Dominant and recessive phenotypes.
(1) Parental generation.
(2) F_1 generation.
(3) F_2 generation. Dominant (red) and recessive (white) phenotype look alike in the F_1 (first) generation and show a 3:1 ratio in the F_2 (second) generation

Figure2: The genotypes of two independent traits show a 9:3:1 ratio in the F_2 generation. In this example, coat color is indicated by 'B' (brown, dominant) or 'b' (white) while tail length are homozygous for each trait ('SSbb' and 'ssBB'), their children in the F_1 generation are heterozygous at both loci and only show the dominant phenotypes. If the children mate with each other, in the F_2 generation all combination of coat color and tail length occur: 9 are brown/short (purple boxes), 3 are white/short (pink boxes), 3 are brown/long (blue boxes) and 1 is white/long (green box).

Figure3: The color alleles of *Mirabilis jalapa* are not dominant or recessive
(1) Parental generation.
(2) F_1 generation.
(3) F_2 generation. The 'red' and 'white' allele together make a 'pink' phenotype, resulting in a 1:2:1 ratio of red:pink:white in the F_2 generation.

The reason for these laws is found in the nature of cell nucleus. It is made up of several chromosomes carrying the genetic traits. In a normal cell, each of these chromosomes has two parts, the chromatids. A reproductive cell, which is created in meiosis, usually contains only one of those chromatids of each chromosome.

By merging two of these cells (usually one male and one female), the full set is restored and the genes are mixed. The resulting cell becomes a new embryo. The fact that this new life has half the genes of each parent (23 from mother, 23 from father for total of 46) is one reason for the Mendelian laws.

The second most important reason is the varying dominance of different genes, causing some traits to appear unevenly instead of averaging out (whereby dominant doesn't mean more likely to reproduce; recessive genes can become the most common, too).

There are several advantages of this method (sexual reproduction) over reproduction without genetic exchange (asexual reproduction):

1. Instead of nearly identical copies of an organism, a broad range of offspring develops, allowing more different abilities and evolutionary strategies.

2. There are usually some errors in every cell nucleus. Copying the genes usually adds more of them. By distributing them randomly over different chromosomes and mixing the genes, such errors will be distributed unevenly over the different children. Some of them will therefore have only very few such problems. This somewhat helps reduce problems with copying errors.

3. Genes can spread faster from one part of a population to another. This is for instance useful if there's a temporary isolation of two groups. New genes developing in each of the populations don't

get reduced to half when one side replaces the other, they mix and form a population with the advantages of both sides.

4. Sometimes, a mutation (ex. Sickle cell anemia) can have positive side effects (in this case malaria resistance). The mechanism behind the Mendelian laws can make it possible for some offspring to carry the advantages without the disadvantages until further mutations solve the problems.

Mendelian Trait

A Mendelian trait is one that is controlled by a single locus and shows a simple Mendelian inheritance pattern. In such cases, a mutation in a single gene can cause a disease that is inherited according to Mendel's laws. Examples include Sickle-cell anemia, Tay-Sachs disease, cystic fibrosis and xeroderma pigmentosa.

A disease controlled by a single gene contrasts with a multi-factorial disease, like arthritis, which is affected by several loci (and the environment) as well as those diseases inherited in a non-Mendelian fashion. The Mendelian inheritance in man data-base is a catalog of, among other things, genes in which Mendelian mutants causes disease.

In genetics, dominance describes a specific relationship between the effects of different versions of a gene (alleles) on a trait (phenotype). Animals (including humans) and plants are mostly diploid, with two copies of each gene, one inherited from each parent.

If the two copies are not identical (not the same allele), their combined effect may be different than the effect of having two identical copies of a single allele. But if the combined effect is the same as the effect of having two copies of one of the alleles, we say that allele's effect is dominant over the other.

For example, having two copies of one allele of the EYCL3 gene causes the eye's iris to be brown, and having two copies of another allele causes the iris to be blue. But having one copy of each allele leads to a brown iris. Thus the brown allele is said to be dominant over the blue allele (and the blue allele is said to be recessive to the brown allele).

We now know that in most cases a dominance relationship is seen when the recessive allele is defective. In these cases a single copy of the normal allele produces enough of the gene's product to give the same effect as two normal copies, and so the normal allele is described as being dominant to the defective allele. This is the case for the eye color alleles described above, where a single functional copy of the 'brown' allele causes enough melanin to be made in the iris that the eyes appear brown even when paired with the non-melanin- producing 'blue' allele.

Dominance was discovered by Mendel, who introduced the use of uppercase letters to denote dominant alleles and lowercase to denote recessive alleles, as is still commonly used in introductory genetics courses (e.g. 'Bb' for alleles causing brown and blue eyes). Although this usage is convenient it is misleading, because dominance is not a property of an allele considered in isolation but of a relationship between the effects of two alleles. When geneticists loosely refer to a dominant allele or a recessive allele, they mean that the allele is dominant or recessive to the standard allele.

Geneticists often use the term dominance in other contexts, distinguishing between simple or complete dominance as described above, and other relationships. Relationships described as incomplete or partial dominance are usually more accurately described as giving an intermediate or blended phenotype. The relationship described as co-dominance describes a relationship where the distinct phenotypes caused by each allele are both seen when both alleles are present.

Nomenclature

Genes are indicated in short-hand by a combination of one or a few letters:

For example, in cat coat genetics the alleles Mc and mc (for 'mackerel tabby') play a prominent role. Alleles producing dominant traits are denoted by initial capital letters; those that confer recessive traits are written with lowercase letters.

The alleles present in a locus are usually separated by a slash '/'; in the 'Mc' vs 'mc' case, the dominant trait is the 'mackerel-stripe' pattern, and the recessive one the 'classic' or 'oyster tabby' pattern, and thus a classical-pattern tabby cat would carry the alleles 'mc/mc', whereas a mackerel-stripe tabby would be either 'Mc/mc' or 'Mc/Mc'.

Relationship to Other Genetics Concepts

Humans have 23 homologous chromosome pairs (22 pairs of autosomal chromosomes and two distinct sex chromosomes, X and Y). It is estimated that the human genome contains 20,000-25,000 genes. Each chromosomal pair has the same genes, although it is generally unlikely that homologous genes from each parent will be identical in sequence.

The Specific Variations Possible for a Single Gene are Called Alleles

For a single eye-color gene, there may be a blue eye allele, a brown eye allele, a green eye allele, etc. Consequently, a child may inherit a blue eye allele from the mother and a brown eye allele from the father. The dominance relationships between the alleles control which traits are and are not expressed.

An example of an autosomal dominant human disorder is Huntington's disease, which is a neurological disorder resulting in impaired motor function. The mutant allele results in an abnormal protein, containing large repeats of the amino acid glutamine.

This defective protein is toxic to neural tissue, resulting in the characteristic symptoms of the disease. Hence, one copy suffices to confer the disorder. A list of human traits that follow a simple inheritance pattern can be found in human genetics. Humans have several genetic diseases, often but not always caused by recessive alleles.

Punnett Square

The genetic combinations possible with simple dominance can be expressed by a diagram called a 'Punnett square'. One parent's alleles are listed across the top and the other parent's alleles are listed down the left side. The interior squares represent possible offspring, in the ratio of their statistical probability. In an example of flower color, 'P' represents the 'dominant purple-colored allele' and 'p' the 'recessive white-colored allele'.

If both parents are purple-colored and heterozygous (Pp), the Punnett square for their offspring would be:

	P	p
P	PP	Pp
p	Pp	pp

In the PP and Pp cases, the offspring is purple colored due to the dominant P. Only in the pp case is there expression of the recessive white-colored phenotype. Therefore, the phenotypic ratio in this case is 3:1, meaning that F2 generation offspring will be purple-colored three times out of four, on average. Dominant alleles are capitalized.

Dominant Allele

Dominant trait refers to a genetic feature that hides the recessive trait in the phenotype of an individual. A dominant trait is a phenotype that is seen in both the homozygous 'AA' and heterozygous 'Aa' genotypes. Many traits are determined by pairs of complementary genes, each inherited from a single parent.

Often when these are paired and compared, one allele (the dominant) will be found to effectively shut out the instructions from the other, the recessive allele. For example, if a person has one allele for blood type A and one for blood type O, that person will always have blood type A. For a person to have blood type O, both their alleles must be O (recessive).

When an individual has two dominant alleles (AA), the individual is referred to as homozygous dominant; an individual with two recessive alleles (aa) is called homozygous recessive. An individual carrying one dominant and one recessive allele is referred to as heterozygous. A dominant trait when written in a genotype is always written before the recessive gene in a heterozygous pair. A heterozygous genotype is written Aa, not aA.

Types of Dominances

Consider the simple example of flower color in peas. The dominant allele is purple and the recessive allele is white.

In a given individual, the two corresponding alleles of the chromosome pair fall into one of the three patterns:

i. Both alleles purple (PP)

ii. Both alleles white (pp)

iii. One allele purple and one allele white (Pp)

If the two alleles are the same (homozygous), the trait they represent will be expressed. But if the individual carries one of each allele (heterozygous), only the dominant one will be expressed. The recessive allele will simply be suppressed.

Simple Dominance in Pedigrees

Dominant traits are recognizable by the fact that they do not skip generations, as recessive traits do. It is therefore quite possible for two parents with purple flowers to have white flowers among their progeny, but two such white offspring's could not have purple offspring (although very rarely, one might be produced by mutation). In this situation, the purple individuals in the first generation must have both been heterozygous (carrying one copy of each allele).

Incomplete Dominance

Discovered by Karl Correns, incomplete dominance (sometimes called partial dominance) is a heterozygous genotype that creates an intermediate phenotype. In this case, only one allele (usually the wild type) at the single locus is expressed in a dosage dependent manner, which results in an intermediate phenotype.

A cross of two intermediate phenotypes (monohybrid heterozygotes) will result in the reappearance of both parent phenotypes and the intermediate phenotype. There is a 1:2:1 phenotype ratio instead of the 3:1 phenotype ratio found when one allele is dominant and the other is recessive. This lets an organism's genotype be diagnosed from its phenotype without time-consuming breeding tests. The classic example of this is the color of carnations.

	R	R'
R	RR	RR'
R'	RR'	R'R'

R is the allele for red pigment. R' is the allele for no pigment. Thus, RR offspring make a lot of red pigment and appear red. R'R' offspring make no red pigment and appear white. Both RR' and R'R offspring make some pigment and therefore appear pink.

Co-dominance

In co-dominance, neither phenotype is recessive. Instead, the heterozygous individual expresses both phenotypes. A common example is the ABO blood group system.

The gene for blood types has three alleles:

A, B, and i. i causes O type and is recessive to both A and B. The A and B alleles are co-dominant with each other. When a person has both an A and a B allele, the person has type AB blood. When two persons with AB blood type have children, the children can be type A, type B, or type AB.

There is a 1A:2AB:1B phenotype ratio instead of the 3:1 phenotype ratio found when one allele is dominant and the other is recessive. This is the same phenotype ratio found in mattings of two organisms that are heterozygous for incomplete dominant alleles. Ex. Punnett square for a father with A and i, and a mother with B and i-

	A	i
B	AB	B
i	A	O

Dominant Negative

Some gain-of-function mutations are dominant and are called 'dominant negative' or antimorphic mutations. Typically, a dominant negative mutation occurs when the gene product adversely affects the normal, wild-type gene product within the same cell. This usually occurs if the product can still interact with the same elements as the wild-type product, but block some aspect of its function. Such proteins may be competitive inhibitors of the normal protein functions.

Types

1. A mutation in a transcription factor that removes the activation domain, but still contains the DNA binding domain. This product can then block the wild-type transcription factor from binding the DNA site leading to reduced levels of gene activation.
2. A protein that is functional as a dimer. A mutation that removes the functional domain, but retains the dimerization domain would cause a dominant negative phenotype, because some fraction of protein dimers would be missing one of the functional domains.

Autosomal Dominant Gene

Autosomal Dominant Pedigree Chart

An autosomal dominant gene is one that occurs on an autosomal (non-sex determining) chromosome. As it is dominant, the phenotype it gives will be expressed even if the gene is heterozygous. This contrasts with recessive genes, which need to be homozygous to be expressed.

The chances of an autosomal dominant disorder being inherited are 50% if one parent is heterozygous for the mutant gene and the other is homozygous for the normal, or 'wild-type', gene. This is because the offspring will always inherit a normal gene from the parent carrying the wild-type genes, and will have a 50% chance of inheriting the mutant gene from the other parent.

If the mutant gene is inherited, the offspring will be heterozygous for the mutant gene, and will suffer from the disorder. If the parent with the disorder is homozygous for the gene, the offspring produced from mating with an unaffected parent will always have the disorder.

The term vertical transmission refers to the concept that autosomal dominant disorders are inherited through generations. This is obvious when you examine the pedigree chart of a family for a particular trait. Because males and females are equally affected, they are equally likely to have affected the children. Although the mutated gene should be present in successive generations in which there are more than one or two offspring, it may appear that a generation is skipped if there is reduced penetrance.

Autosomal Dominant Disorders

1. Achondroplasia
2. Anti-thrombin deficiency
3. Autosomal dominant polycystic kidney disease, ADPKD (Adult-onset)
4. BRCA1 and BRCA2 mutations (Hereditary breast ovarian cancer syndrome)
5. Brugada syndrome
6. Charcot-Marie-Tooth syndrome
7. Ectrodactyly
8. Cleft Chin
9. Ehlers-Danlos syndrome
10 Familial hypercholesterolemia
11. Familial adenomatous polyposis
12. Facioscapulohumeral muscular dystrophy
13. Fatal familial insomnia
14. Fibrodysplasia ossificans progressiva
15. Hereditary hemorrhagic telangiectasia (Osler-Weber-Rendu syndrome)
16. Hereditary spherocytosis
17. Hunting's disease
18. Hypertrophic cardiomyopathy
19. Kabuki syndrome (potentially)
20. Lactase persistence
21. Malignant hyperthermia
22. Mandibulofacial dysostosis
23. Marfan syndrome
24. Mowat-Wilson syndrome
25. Multiple endocrine neoplasia

26. Noonan syndrome
27. Neurofibromatosis
28. Osteogenesis imperfect
29. Pfeiffer syndrome
30. Tuberous sclerosis

Recessive Allele

The term 'recessive allele' refers to an allele that causes a phenotype (visible or detectable characteristic) that is only seen in homozygous genotypes (organisms that have two copies of the same allele) and never in heterozygous genotypes.

Every diploid organism, including humans, has two copies of every gene on autosomal chromosomes, one from the mother and one from the father. The dominant allele of a gene will always be expressed while the recessive allele of a gene will be expressed only if the organism has two recessive forms. Thus, if both parents are carriers of a recessive trait, there is a 25% chance with each child to show the recessive trait.

The term 'recessive allele' is part of the laws of Mendelian inheritance formulated by Gregor Mendel. Examples of recessive traits in Mendel's famous pea plant experiments include the color and shape of seed pods and plant height.

Autosomal Recessive Allele

Autosomal recessive is a mode of inheritance of genetic traits located on the autosomes (the pairs of non-sex determining chromosomes—22 in humans).

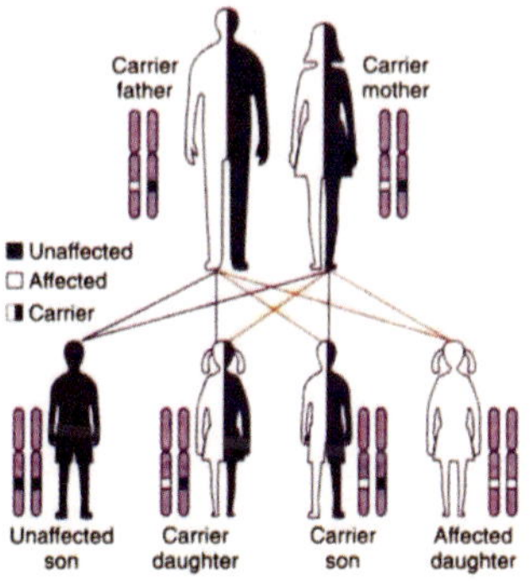

Relationship between two carrier parents and probabilities of children being unaffected, carriers, or affected

In opposition to autosomal dominant trait, a recessive trait only becomes phenotypically apparent when two similar alleles of a gene are present. In other words, the subject is homozygous for the trait.

The frequency of the carrier state can be calculated by the Hardy-Weinberg formula:

$P^2 + 2pq + q^2 = 1$ (p is the frequency of one pair of alleles, and $q = 1 - p$ is the frequency of the other pair of alleles).

Recessive genetic disorders occur when both parents are carriers and each contributes an allele to the embryo, meaning these are not dominant genes. As both parents are heterozygous for the disorder, the chance of two disease alleles landing in one of their offspring is 25% (in autosomal dominant traits this is higher).

50% of the children (or 2/3 of the remaining ones) are carriers. When one of the parents is homozygous, the trait will only show in his/her offspring if the other parent is also a carrier. In that case, the chance of disease in the offspring is 50%.

Autosomal Recessive Disorders

Dominance/recessiveness refers to phenotype, not genotype. An example to prove the point is sickle cell anemia. The sickle cell genotype is caused by a single base pair change in the beta-globin gene, normal is GAG (glu) and sickle is GTG (val).

There are several phenotypes associated with the sickle genotype:

1. Anemia (a recessive trait)
2. Blood cell sickling (co-dominant)
3. Altered beta-globin electrophoretic mobility (co-dominant)
4. Resistance to malaria (dominant)

This example demonstrates that one can only refer to dominance/recessiveness with respect to individual phenotypes.

Other Recessive Disorders

1. Albinism
2. lpha 1-antitrypsin deficiency
3. Autosomal Recessive Polycystic Kidney
4. loom's syndrome
5. Certain forms of spinal muscular atrophy
6. Chronic granulomatous disease
7. Congenital adrenal hyperplasia deficiency
8. Cystic fibrosis
9. Dry (also known as 'rice-bran') earwax
10. Dubin-Johnson syndrome
11. Familial Mediterranean fever Disease-ARPKD (Child-onset)
12. Fanconi anemia
13. Friedreich's ataxia

14. Galactosemia
15. GluCoSC-6-phoSphate dehydrogenase
16. Glycogen storage diseases
17. Haemochromatosis types 1-3
18. Homocystinuria
19. Mucopolysaccharidoses
20. Pendred syndrome
21. Phenylketonuria
22. Rotor syndrome
23. Tay-Sach's disease
24. Thalassemia
25. Wilson's disease
26. Xeroderma pigmentosum

Mechanisms of Dominance

Many genes code for enzymes. Consider the case where someone is homozygous for some trait. Both alleles code for the same enzyme, which causes a trait. Only a small amount of that enzyme may be necessary for a given phenotype.

The individual therefore has a surplus of the necessary enzyme. Let's call this case 'normal'. Individuals without any functional copies cannot produce the enzyme at all, and their phenotype reflects that. Consider a heterozygous individual. Since only a small amount of the normal enzyme is needed, there is still enough enzyme to show the phenotype. This is why some alleles are dominant over others.

In the case of incomplete dominance, the single dominant allele does not produce enough enzymes, so the heterozygotes show some different phenotype. For example, fruit color in eggplants is inherited in this manner.

A purple color is caused by two functional copies of the enzyme, with a white color resulting from two non-functional copies. With only one functional copy, there is not enough purple pigment, and the color of the fruit is a lighter shade, called violet.

Some non-normal alleles can be dominant. The mechanisms for this are varied, but one simple example is when the functional enzyme E is composed of several subunits where each Ei is made of several alleles Ei = ailai2, making them either functional or not functional.

For example one could have the rule that if any of the Ei subunits are non-functional, the entire enzyme E is non-functional in the sense that the phenotype is not displayed. In the case of a single subunit say El is El = F where F has a functional and non-functional allele (heterozygous individual) (F = a1A1), the concentration of functional enzyme determined by E could be 50% of normal. If the enzyme has two identical subunits (the concentration of functional enzyme is 25% of normal).

For four subunits, the concentration of functional enzyme is about 6% of normal (roughly scaling slower than 1/2c where c is the number of copies of the allele-1/24 is about 51% percent). This may not be enough to produce the wild type phenotype. There are other mechanisms for dominant mutants.

Sex-limited Genes

Sex-limited genes are genes which are present in both sexes of sexually reproducing species but turned on in only one sex. In other words, sex-limited genes cause the two sexes to show different traits or phenotypes. An example of sex-limited genes are genes which instructs male elephant seal to grow big and fight, at the same time instructing female seals to grow small and avoid fights. These genes are responsible for sexual dimorphism.

Gene Pool

A gene pool is the collection of different genes within an interbreeding population. The concept of a gene pool usually refers to the sum of all the alleles at all of the loci within the genes of a population of a single species. It includes both genes that are expressed, and those that are not.

A population that has a large and diverse gene pool tends have increased biological fitness and is usually able to adapt to and survive throughout pressures caused by rapid changes in environmental conditions or disease. Although individuals may die in such situations, a diverse population often contains enough genetic variation so that there will be an availability of suitable genes that are necessary for survival. This is the basic concept behind the theory of natural selection.

A population with a narrow gene pool containing low diversity is more likely to suffer from reduced fitness when affected by stresses and is more likely to become extinct. This is either due to a low availability of alleles that are necessary for an organism to survive under certain conditions, or due to the consequences of inbreeding. When there is less variation between available alleles, there is a greater probability that a harmful allele, which causes a deformity or disease, will increase in frequency.

Over time, the size of any gene pool can change through natural selection, gene flow and genetic drift. The size may increase when a gene mutation occurs and survives, and decrease when alleles are removed, e.g. when a population goes through a bottleneck.

There can be several variations of each allele for each gene, some of which are dominant alleles and some of which are recessive alleles. Within genes, each allele variant occurs at a particular frequency, at a particular time. When small changes to allele frequency occur this is known as microevolution, large changes, or an accumulation of small changes results in macroevolution, both of which can result in speciation.

The genetic variation within the population is characterized by differences in allele frequency and determines the relative frequency of each phenotype displayed in a population. The phenotype of an individual is determined by its own genotype.

Examples of Gene Pool

Humans

Every human being on Earth is able to interbreed with one another as a single species. The human gene pool is therefore made up of every allele variant of the approximated 19,000-20,000 human genes within our DNA.

Butterflies

In a population of butterflies, there are two alleles at locus one. Individuals with the dominant allele have eyespots on their wings, while individuals whom are homozygous (have two of the same copies) for the recessive allele do not have eyespots.

A new predator is introduced, which rather than being deterred, is attracted to the eyespots.

The predator soon feeds on all of the butterflies that have eyespots. Because the eyespot allele is dominant, it is always expressed. So by predating all butterflies with an eyespot, the predator removes the allele for eyespots and the size of the butterfly gene pool decreases.

Eyespots could return to the population if a mutation of an allele occurs which causes eyespots and it is favored by natural selection. Alternatively, eyespots could return through gene flow if individuals from the original butterfly population mate with another butterfly population that has the allele for eyespots.

The image shows a butterfly with eyespots on its wing. In our example, the butterflies with eyespots are predated, which removes the allele that causes the eyespots, leaving only the recessive gene.

Speciation

Speciation is the formation of new and distinct species in the course of evolution. Speciation involves the splitting of a single evolutionary lineage into two or more genetically independent lineages.

In eukaryotic species—that is, those whose cells possess a clearly defined nucleus—two important processes occur during speciation: the splitting up of one gene pool into two or more separated gene pools (genetic separation) and the diversification of an array of observable physical characteristics (phenotypic differentiation) in a population. There are many hypotheses about how speciation starts, and they differ mainly in the role of geographic isolation and the origin of reproductive isolation (the prevention of two populations or more from interbreeding with one another).

Allopatric Speciation

Geographic isolation most often occurs with populations that are completely separated (allopatry) by a physical barrier, such as a mountain range, river, or desert. The separated populations adapt to their own unique environments, becoming so genetically different from one another that members of one population cannot breed with members of the other. Examples of allopatric speciation abound, and the process is considered the dominant form of speciation in organisms that engage in sexual reproduction. Darwin's finches on the Galapagos Islands, which may have speciated allopatrically because of volcanic eruptions that divided populations, is a famous example.

Sympatric Speciation

A controversial alternative to allopatric speciation is sympatric speciation, in which reproductive isolation occurs within a single population without geographic isolation. In general, when populations are physically separated, some reproductive isolation arises. How genetic divergence can happen within a population of individuals that are continually interacting with one another is usually difficult to explain. Examples of sympatric speciation are often disputed because they must show convincing evidence of species descending from the same ancestral species, the reproductive isolation of the group, and of allopatry not causing the speciation.

Nevertheless, sympatric speciation has been shown to have occurred in apple maggot flies (*Rhagoletis pomonella*), a parasitic insect that laid its eggs in the fruit of wild hawthorns (*Crataegus*) until one subset of the population began to lay its eggs in the fruit of domesticated apple trees (*Malus domestica*) that grew in the same area. That small group of apple maggot flies selected a different host species from the rest of its kind, and its offspring became accustomed to domesticated apples and later laid their own eggs in them, thereby cementing the shift in host.

Sympatric selection might also result from a combination of sexual selection and ecological factors. Studies of African cichlid fishes in Lake Nyasa and other lakes in the East African Rift System record so-called species flocks (individuals of the same species that "flock" together in one large assemblage) that have arisen in ecologically uniform lakes. Such a condition substantially reduces the chances of allopatry being the cause of speciation, and it may result in groups of females within a population developing a strong affinity for males with different extreme phenotypic traits, such as scale markings and limbs that differ in size from average individuals. Other studies suggest that sympatry among cichlid fishes also occurs in rivers feeding the East African Rift System lakes, as well as in Nicaragua's crater lakes.

Cichlid fish (family Cichlidae).

Genetic Drift

Genetic drift is a change in allele frequency in a population, due to the forces of chance. Oftentimes, mutations in cells can have no effect. These changes in genetics can increase or decrease in a population, simply due to chance. Although variations of genes (also known as alleles) can be selected for because they help or hinder an organism with reproducing. When the allele is not responsible for the change in its frequency in a population, genetic drift is acting on the allele.

Genetic drift is much more likely in smaller populations of organisms. The individual lines in the graph above track the frequency of alleles in a given population. When the population is small, and many alleles exist, any of the alleles can quickly become fixed or extinct in the population. When there are many organisms in the population, there is less of a chance of losing an entire allele, because many organisms carry the allele and it is less likely they will all be wiped out.

Genetic drift can easily be confused with natural selection. The difference is whether or not the allele is actively participating in the change in allele frequencies. If the allele affects an organism in a way that causes more reproduction of the DNA, the allele will increase in frequency. If it causes harm, it will decrease. This is caused by the allele's direct effects on the organism and the environment. This is natural selection. When the allele is increased or decreased simply because it was present in the organisms that survived, this is genetic drift.

Types of Genetic Drift

Population Bottleneck

A population bottleneck is a type of genetic drift in which a population's size severely decreases. This could be caused by predators, disease, or competition from another source. As the population becomes smaller, certain alleles change in frequency in the population simply because the organisms that carry them get killed or eaten. The other alleles increase in frequency, simply because they are the only alleles left. This type of genetic drift can be seen when people don't take their entire course of antibiotics.

Antibiotics kill harmful bacteria in your system, regardless of what alleles they have. The populations of harmful bacterial that cause symptoms are severely reduced until they no longer cause symptoms. If someone quits their antibiotic regiment before all the bacteria are killed, a small population is left. This much smaller population could have allele frequencies that are very different from the original population of bacteria. These changes do not reflect the success or failure of the different alleles, but rather the effects of random selection of bacteria. The new alleles will dominate the population until selection or more genetic drift cause the allele frequencies to change.

Founder Effect

In another type of genetic drift known as the founder effect, a new population is formed, or founded, in a new location. If this new population does not interact and reproduce with the main population, the allele frequencies in this population will be much different from that of the parent population. Many island contain species that only exist on a single island because of the founder effect. For instance, if only two birds of a species land on an island, their alleles alone will account for the diversity present. While these alleles will dominate at first, mutations will arise in the population that will lead to new adaptations. Because the two populations of birds are separated, they cannot share this new adaptation. With enough time, the two populations can diverge to a point which they can no longer interbreed. This is one way that separate species are created.

Examples of Genetic Drift

In a Hypothetical Population

A population of rabbits lives in the woods. The rabbits have many different coat colors: black, brown, tan, white, grey, and even red. In the population, the different alleles that create coat color are equally distributed. A disease comes into the rabbit population and kills 90% of the rabbits. The only rabbits that are left are red and grey rabbits, simply by chance. The genes have thus "drifted" from 6 alleles to only 2. This is an example of a bottleneck effect.

In Real Life

Genetic drift happens all the time in populations, although it is not easily seen. Often, mutations arise that have little effect on the organism. These mutations get passed on if the organism reproduces, and do not get passed on if the organism does not survive. Although genetic drift used to be thought of in only small populations, even large populations experience genetic drift of certain alleles, because a small number of individuals carry the alleles. Whether or not these alleles are duplicated is not a function of natural selection, but of chance. Many alleles come or go in populations without affecting great change.

Some Other Examples

- Of the two pink monkeys in the world – one male, one female – the female dies, ensuring that there will never be a pure-bred pink monkey again.
- A random succession of births results in all other hair colors going extinct within a village full of redheaded people.
- The freckled dazzle-flower exchanges genetic material with a small population of un-freckled dazzle-flowers. After a few seasons, there are no un-freckled dazzle-flowers left.
- The last green-eyed person in a small town dies, leaving only brown-eyed and blue-eyed people.
- An airplane crash introduces the white heron, which loves spotted mackerel, into a population of spotted and unspotted mackerel. Over time, fewer mackerel are born with spots.
- A man steps on a group of beetles, randomly killing most of the green ones but leaving most of the brown ones alive, resulting in fewer green beetles being produced in the population.
- A wildflower population consisting of blue, purple, and pink flowers is subjected to a mudslide that kills most of the blue ones. As time progresses, blue flowers eventually die out, leaving only purple and pink wildflowers.
- Due to random successions of births, a town has an unusually high population of people with strawberry blonde hair, a trait that increases over time and leaves very few people with different hair colors.
- A large population of marmots, about half of which have spots, becomes very ill. More spotted marmots than un-spotted marmots are left; as a result, un-spotted marmots become very rare as time progresses.
- The very last red-billed mallard dies, leaving only orange-billed mallard populations behind.
- A large group of red poppies exchanges genetic material with white poppies. Over time, white poppies become rarer and rarer while red poppies thrive.

References

- Dna-types-structure-and-function-of-dna-1301: biologydiscussion.com, Retrieved 25 May 2018
- RNA, science, britannica.com: Retrieved 09 July 2018
- Chromosome: biologydictionary.net, Retrieved 09 July 2018
- Meiosis, cell-cycle-and-cell-division: toppr.com, Retrieved 25 May 2018
- Phases-of-meiosis, cellular-molecular-biology, biology, science: khanacademy.org, Retrieved 10 July 2018
- Biological-inheritance-meaning-categories-specifications-and-mendels-laws-17461: biologydiscussion.com, Retrieved 25 May 2018
- Speciation, science: britannica.com, Retrieved 09 May 2018

4

Evolution

Any change expressed in the hereditable traits of organisms over successive generations is termed as evolution. Such changes are mediated by genes when they are inherited from parent to offspring during the process of reproduction. The aim of this chapter is to provide an insight into the phenomenon of evolution through the elucidation of topics such as genetic variation, mutation and Darwin's theory.

In biology, evolution is the change in the characteristics of a species over several generations and relies on the process of natural selection.

The theory of evolution is based on the idea that all species are related and gradually change over time.Evolution relies on there being genetic variation in a population which affects the physical characteristics (phenotype) of an organism.Some of these characteristics may give the individual an advantage over other individuals which they can then pass on to their offspring.

Natural Selection

Charles Darwin's theory of evolution states that evolution happens by natural selection.

Individuals in a species show variation in physical characteristics. This variation is because of differences in their genes.Individuals with characteristics best suited to their environment are more likely to survive, finding food, avoiding predators and resisting disease. These individuals are more likely to reproduce and pass their genes on to their children.

Individuals that are poorly adapted to their environment are less likely to survive and reproduce. Therefore their genes are less likely to be passed on to the next generation.

As a consequence those individuals most suited to their environment survive and, given enough time, the species will gradually evolve.

Natural Selection in Action: the Peppered Moth

Before the industrial revolution in the mid-1700s, the peppered moth was most commonly a pale whitish colour with black spots.This colouring enabled them to hide from potential predators on trees with pale-coloured bark, such as birch trees.The rarer dark-coloured peppered moths were easily seen against the pale bark of trees and therefore more easily seen by predators.

A pale peppered moth on an oak tree.

As the Industrial Revolution reached its peak, the air in industrial areas became full of soot. This stained trees and buildings black.As a result, the lighter moths became much easier to spot than the darker ones, making them vulnerable to being eaten by birds.The darker moths were now camouflaged against the soot-stained trees and therefore less likely to be eaten.Over time this change in the environment led to the darker moths becoming more common and the pale moths rarer.

A pale peppered moth on a dark tree.

Role of Genes

The mechanisms of evolution operate at the genomic level. Changes in DNA sequences affect the composition and expression of our genes, the basic units of inheritance.To understand how different species have evolved we have to look at the DNA sequences in their genomes. Our evolutionary history is written into our genome. The human genome looks the way it does because of all the genetic changes that affected our ancestors. When DNA and genes in different species look very similar, this is usually taken as evidence of them sharing ancestors. For example, humans and the fruit fly, Drosophila melanogaster, share much of their DNA. 75 per cent of genes that cause diseases in humans are also found in the fruit fly. DNA accumulates changes over time. Some of these changes can be beneficial, and provide a selective advantage for an organism. Other changes may be harmful if they affect an important, everyday function. As a result some genes do not change much. They are said to be conserved.

Different Types of Evolution

Convergent Evolution

When the same adaptations evolve independently, under similar selection pressures. For example, flying insects, birds and bats have all evolved the ability to fly, but independently of each other.

Co-evolution

When two species or groups of species have evolved alongside each other where one adapts to changes in the other.For example, flowering plants and pollinating insects such as bees.

Adaptive Radiation

When a species splits into a number of new forms when a change in the environment makes new resources available or creates new environmental challenges. For example, finches on the Galapagos Islands have developed different shaped beaks to take advantage of the different kinds of food available on different islands.

Sketches of the heads of finches from the Galapagos Islands showing the differences in their beak shapes due to evolution.

Genetic Variation

Genetic variation is a term used to describe the variation in the DNA sequence in each of our genomes. Genetic variation is what makes us all unique, whether in terms of hair colour, skin colour or even the shape of our faces.

Individuals of a species have similar characteristics but they are rarely identical, the difference between them is called variation. Genetic variation is a result of subtle differences in our DNA. Single nucleotide polymorphisms (SNPs, pronounced 'snips') are the most common type of genetic variation amongst people. Each single nucleotide polymorphism represents a difference in a single DNA base, A, C, G or T, in a person's DNA. On average they occur once in every 300 bases and are often found in the DNA between genes. Genetic variation results in different forms, or alleles, of genes. For example, if we look at eye colour, people with blue eyes have one allele of the gene for eye colour, whereas people with brown eyes will have a different allele of the gene. Eye colour, skin tone and face shape are all determined by our genes so any variation that occurs will be due to the genes inherited from our parents. In contrast, although weight is partly influenced by our genetics, it is strongly influenced by our environment. For example, how much we eat and how often we exercise. Genetic variation can also explain some differences in disease susceptibility and how people react to drugs. Genetic variation is important in evolution. Evolution relies on genetic variation that is passed down from one generation to the next. Favourable characteristics are 'selected' for, survive and are passed on. This is known as natural selection.

Diversity of Life

The Convention on Biological Diversity defines biodiversity as 'the variability among living organisms from all sources including, inter alia, terrestrial, marine and other aquatic ecosystems and the ecological complexes of which they are part; this includes diversity within species, between species and of ecosystems'. According to this definition, three main levels of biological diversity can be identified: genetic diversity, species diversity and ecosystem diversity.

Genetic diversity is the heritable variation within a single species, including the differences among individuals in a local population. In an evolutionary perspective, this is the ultimate source of all kinds of diversity in the biosphere. On the other hand, ecosystem diversity is possibly the level of biodiversity most obvious to the lay observer, because of the immediate visual impact of the differences among aquatic and terrestrial landscapes or vegetation types, such as a pond and a seashore, a conifer forest and an alpine meadow. However, its measurement suffers from major problems of standardization. Most approaches to biological diversity therefore focus on species (or taxonomic) diversity.

A first-level explanation of the diversity of life on earth is the diversity of earth itself. There are two major aspects of geographical diversity of the physical environment that allow living beings to become numerous. One aspect is habitat heterogeneity at local, regional, continental and even global scale. Organisms successfully thriving in a wide spectrum of different habitats are rare, the bulk of living species being instead confined, more or less strictly, to a narrow set of environmental conditions. The physical heterogeneity of the planet's surface, however, does not explain why similar habitats in different continents, and even in different regions within the same continent, are inhabited by widely dissimilar species. This is explained, instead, by history. Physical or ecological barriers between similar habitat patches may interrupt gene flow to such an extent as to bring about allopatric speciation. Similar habitats in individual islands within an archipelago or on individual peaks within a rugged mountain range are commonly inhabited by related but different species of sedentary animals, such as land snails or wingless beetles. All these species, whose geographic range may be restricted to a few square kilometres, arose because of the physical or ecological barriers that interrupted the genetic flow between populations.

In oceanic archipelagos, this condition of geographical isolation may affect the whole biota. For instance, the native fauna and flora of oceanic islands such as the Hawaiian chain is to a very large extent endemic: 89% for angiosperms, 99% for insects. Moreover, within this single island chain a large number of species are confined to one single island, or even to a single district within one island ,due to the habitat fragmentation caused by local topography or by recent lava flows. It has been estimated that the more than 10 000 animal and plant species now inhabiting the Hawaiian archipelago evolved there from a few hundred successful colonizers, most of them of North American origin.

The importance of geographical isolation in determining high levels of species diversity is also apparent in a comparison between marine and freshwater fishes. Of all fish species described to date (some 25 000), those living in the sea are less than twice as numerous as those living in inland waters, although the total volume of oceanic waters is about ten thousand times greater than the volume of inland waters. The relatively enormous diversity of fresh-water fish species is explained by the fact that inland waters are fragmented into thousands of more or less completely isolated basins, a condition largely facilitating allopatricspeciation.

In more general terms, it has been estimated that only 15% of all living species described to date inhabit the sea. It is unlikely that future investigations will significantly alter this ratio. Likely explanations of this unbalanced distribution of diversity include the higher heterogeneity of continental environments and their higher structural(architectural) complexity with respect to the conditions prevailing in the oceans. At a higher taxonomic level, however, animal life is more diverse in the sea than on land. All animal phyla are represented in the sea and several phyla (e.g. echinoderms, ctenophores, sipunculans, brachiopods) are exclusively marine. This may be due, in part, to the fact that life originated in the sea and remained confined to this realm during much of its history;

no less important, however, is the effect of the strict adaptations required for living in terrestrial (and, to a lesser degree, freshwater) environments, adaptations that cannot be met by animals with body designs like those of a sea urchin or a jelly fish.

A second major explanation of the diversity of life is found in the multiple adaptations developed by most living beings in relation to the other organisms with which they interact, be these competitors, prey, predators, hosts or symbionts. The relevance of these interactions for the evolution of biological diversity is particularly conspicuous when two species interact so closely that each of them represents a major selective agent in the evolution of the other, thus offering a case for coevolution. Interesting examples of coevolution are found in the relationships between flowering plants and their insect pollinators: in many instances, the two partners are so closely specialized that size and shape of the insect's mouthparts, temporal flight schedules etc. are strictly matched by the shape of the corolla, the location of the nectaries, the length and shape of the stamens, and the timing of flower opening. Large plant families such as orchids (around 18 000 species) and legumes (around 16 500 species), and large genera such as Ficus (around 800 species), owe much of their conspicuous species richness to their strict interactions with specialized pollinators.

Interspecific relationships are also crucial in explaining the astonishing diversity found in several groups of parasites. Most parasites attack a very restricted number of host species, sometimes just one. This explains, for example, the remarkable diversity found in Eimeria,agenus of sporozoan protists: more than one thousand species have been described to date and it has been estimated that in this genus there may exist some 35 000species, each of them attacking a selected group (mostly agenus, or even a single species) of vertebrate (rarely invertebrate) hosts. The same will possibly apply to other groups of parasites, e.g. to several families of nematodes.

A similar degree of host specificity is often found in small animals (many groups of insects and mites) and fungi(especially rusts, Puccinia and relatives) living on flowering plants. Adaptations to their hosts involve these parasites 'specializations to exploit only selected parts of the host plant, e.g. young leaves, mature leaves, stem, roots, seeds etc., so that many dozen parasite species may attack the same host without directly interacting with each other.

We must acknowledge, however, that neither geographical isolation nor specific adaptations to other organisms can help explain those extraordinary instances of biological diversity that are commonly known as species flocks. These are groups of dozens and even hundreds of species, all clearly derived from a single common ancestor from which they diversified without perceptible geographical isolation and now all living in the closest geographical proximity within a restricted geographical area. The most species-rich and best investigated species flocks are those of the African cichlid fishes living in three large fresh water basins, Lake Victoria, Lake Tanganyika and Lake Malawi. Each of these lakes hosts a few hundred different cichlid species, each of them with its strikingly different morphological, ecological and behavioral adaptations, living alongside its closest relatives inhabiting the same lake. In the case of Lake Victoria, the cichlid species flock developed from a single ancestor since the last dramatic desiccation of the whole basin dated c.12 000 years ago, a very short time span to account for the origin of some three hundred species from a single ancestor.

As insects represent more than one half of the total biological diversity on Earth, it is sensible to ask the question, why are insects so numerous? A first explanation of their unique diversity is to be found in their size. Insects cannot be more than a few centimeters across, due to structural constraints such as the mechanical properties of their exoskeleton and the effciency of gas diffusion in their tracheal system; on the other hand, their complex architecture cannot be easily accommodated in much less than 1 mm length. In fact, most insects are between 1 and 20 mm long. Insects not being individually too big, do not require large areas for populations to establish themselves, therefore no long-distance displacements are generally necessary for either feeding or reproduction. On the other hand, most insects are either too heavy or too fragile for long-distance passive transport. These conditions facilitate the establishment of isolated populations, a prerequisite for allopatric speciation. A second major cause of insect diversity is their feeding specialization. This is true for phytophagous species as well as for those living as parasitoids of other arthropods. In both cases, high degrees of host specificity are quite common.

Estimates of Current Diversity

The single most commonly used descriptor of species diversity is species number. This number is correlated ,indeed, with some measures of ecological diversity, such as the complexity of food webs, or topographic diversity. However, species richness is, at best, just a measure of one aspect of the global diversity of life.

To improve the information content of biodiversity estimates, it has been suggested that we need to incorporate measures of phylogenetic relatedness among the species present in a given area, so to approach a more informative description of 'character richness' in the sample.

But even estimating species diversity on Earth is not easy. This is due only in part to the incompleteness of our current inventory of biodiversity. There are serious problems, indeed, even with that part of biological diversity that has been already described and named. A first problem arises because of the lack of comprehensive and reliable monographs for many, if not most, of the major groups of living beings. For example, there is no recent world catalogue for popular groups such as beetles(Coleoptera), or butterflies and moths (Lepidoptera): that means that the current estimates of 400 000 described species in the first group and 150 000 in the second may well be some 20% wrong. The major diffculty is not so much retrieving all existing species names from a very scattered literature, but identifying all synonymies; that is, all cases where two or more different names apply to one and the same species. Synonymization requires a critical appraisal of old and new evidence and, as such, requires the time-consuming work of many dedicated specialists. Even for a well-researched group such as the flowering plants, less than 20% of the currently recognized species have been treated in genus- or family-level monographs during the twentieth century.

A more subtle but far from trivial problem derives from the uncertainties in the definition of the basic unit of biodiversity. The circumscription of species may be very different, indeed, if one adopts a biological or a phyloge-netic species concept. This is well exemplified by a 1992study of the birds-of-paradise, a well-known group where traditional classifications, based on the biological species concept, acknowledge the existence of 40 to 42 species, whereas not less than 90 phylogenetic species may bedistinguished in the same group. Still worse is the case of organisms with uniparental reproduction, where the biological species concept simply does not apply, by definition. Examples are offered, in the northern temperate regions, by the brambles (Rubus), the hawthorns (Crataegus) and the dandelions (Taraxacum). In each of these thre egenera, hundreds of species names have been given to slightly different morphotypes, which are perhaps morphologically distinct, but often, occurring together in the same spot, do not behave as different ecological units within the local community. For the strict advocates of the the biological species concept, these living beings simply demonstrate that not all aspects of life are articulated in a biological species; this would imply that describing biodiversity only in terms of species counts is, in principle ,unsatisfactory.

An estimate of the number of species named to date is given in table.

From a geographical point of view, there are some prominent hot spots of biological diversity. For example, the four areas of highest diversity for higher plants are Latin America, where one-third of the world flora is at home with some 85 000 species thus far recorded, China(30 000 species, some 12% of the world total), Mexico(26 000) and Indonesia (20 000).

A latitudinal gradient of biodiversity, with species number decreasing from the Equator to the Poles, is broadly observable despite the existence of many plant and animal groups whose distribution is centred in the temperate areas, such as the Rosaceae, the Cruciferaeand the aphids. These latitudinal differences in species diversity are observed at the local as well as at the region allevel. For instance, on one hectare of tropical forest in Ecuador there may be as much as 473 tree species, whereas in a temperate forest a mere handful of tree species (if not just one species, as in many forest stands in cold temperate areas) may cover hundreds of square kilometres. Historical factors as well as present-day conditions concur to the explanation of the higher species diversity in the tropics. For instance, it has been suggested that during the Pleistocene the Amazonian forest became fragmented into a large number of small areas that acted as refugia for the forest fauna and provided opportunity for intensive allopatric speciation. Later on, when these forest fragments joined together again to form the present-day forest, the species that had differentiated in the separate refugiahad a chance to expand and to become sympatric.

Table: Estimated number of species named to date

	Described ($x10^3$)	Existing (working figure) ($x10^3$)
Bacteria	4	1000
Fungi	75	1000
'Protozoa'	40	300
Álgae'	45	400
Land Plants	270	300
Nematodes	25	500
Crustaceans	45	150
Arachnids	80	750
Insects	1000	10000
Molluscs	100	200
Chordates	50	55
Others	130	300
Total	1900	15000

How much their present coexistence depends on the high complexity of the ecosystem or on its high productivity is still matter for dispute.

The current level of knowledge varies greatly between different groups. In the case of birds, if we disregard the problems arising from adopting different species concepts, we can reasonably expect that no more than a few dozen species remain to be described. In the case of mammals, however, recent descriptions of previously unknown species are not limited to small inconspicuous species of rodents, shrews and bats, but also include, quite un expectedly, some large animals such as the bilkis gazelle (Gazella bilkis) from Yemen, described in 1985, Madagascar's golden lemur (Hapalemur aureus), and four large ruminants from the Vietnamese forests, three of them representing completely new genera, first described in the 1990s: the saola or spindlehorn (Pseudoryx nghetin- ghensis), the linh duong (Pseudonovibos spiralis), a giant muntjac deer (Megamuntiacus vuquangensis) and another muntjac (Muntiacus truongsonensis).

It is for invertebrates, however, that a truly dramatic increase in species description has taken place in the last few decades. An impressive example is provided by arachnids and crustaceans: in these groups, the number of new species described between 1960 and 1970 equals the total number of species described in the same groups during the previous two centuries. This example easily suggests that a large percentage of existing species have not yet been described.

Currently, at least 15 000 species are annually described as new.

Different approaches have been followed to obtain estimates of the number of existing species. These methods generally focus either on less intensively investigated taxonomic groups, such as bacteria, fungi, nematodes, mites and insects, or on some exceptionally species-rich habitats. The two aspects, however, are closely inter- related. For example, insects are the main component of species diversity in the tropical forest canopy, as nematodes are in the deep sea floor.

A sample of arthropods collected on just 10 trees in Borneo included 24 000 specimens, belonging to more than 2800 species. One of the most abundant and diverse groups were the tiny parasitic chalcid wasps: among the 1455 specimens belonging to this group, 739 different species could be counted, 437 of these being represented by just one specimen each. These and similar data have led to an estimation of the total number of arthropod species in the tropical forests worldwide at somewhere between 10 million and 80 million. Much more conservative estimates, however, have been obtained following different approaches. It may be sensible, for instance, to compare the number of described and un described species collected by prolonged sampling efforts in biologically rich and hitherto less investigated areas. Thus, in a very extensive collection of Hemiptera from a topographically diverse area of tropical rainforest in Sulawesi (in Indonesia), the described species amounted to more than one-third of the total, thus suggesting that the total number of extant species of the same group, and of insects at large, would only be round 2.5 million, i.e. less than three times the figure for the species described thus far.

Besides the tropical forest canopy, possibly the richest reservoir of uncharted biodiversity is the deep-sea, despitethe relatively low amount of energy flowing through it. Estimating this aspect of biodiversity, however, is even more problematic than in the case of tropical insects.

Global estimates of existing biodiversity are thus quiteun certain. Figures ranging from 5 to 130 million species have been recently offered for the gross total. Those given in table, although quite subjective, are only slightly highe rthan the working figures most often offered in the literature.

Species counts are the simplest but not the only possible way of describing biodiversity at either local or regional level. Interesting comparisons between ecosystems may be obtained, for instance, by considering the local distributions of species in terms of the average size of adult individuals or in terms of relative or absolute abundance of the species.

Mutation

In biology, mutation is a sudden change in the base pair sequence of the genetic material of a living organism, whether the genetic material be deoxyribonucleic acid (DNA) or ribonucleic acid (RNA). In multicellular organisms that reproduce sexually, mutations can be subdivided into germ line mutations, which can be passed on to descendants, and somatic mutations, which cannot be transmitted to descendants in animals. Plants sometimes can transmit somatic mutations to their descendants asexually or sexually (in case when flower buds develop in somatically mutated part of plant). A new mutation that was not inherited from either parent is called a de novo mutation.

Mutations can be caused by copying errors in the genetic material during cell division, by exposure to ultraviolet or ionizing radiation, chemical mutagens, or viruses, or can occur deliberately under cellular control during processes such as hypermutation.

In evolutionary theory, specifically the theory of evolution by natural selection, mutation is considered the main source of new variation in a population. Although most phenotypic variation is a product of genetic recombination, all new genes are considered to be produced by mutations. For example, a mutation can be postulated for the development of blue eyes in humans. In the two step evolutionary process, production of genetic variation is the chance element, whereas natural selection is not a chance event, but an antichance event. Natural selection is proposed to increase beneficial mutations and select against and eliminate deleterious mutations. However, most mutations actually are harmful, such as seen in genetic diseases, leading to speculation that the role of mutations may be overstated by evolutionary biologists and that other factors, perhaps less random, may be of greater importance in the origin of new designs and macroevolutionary changes.

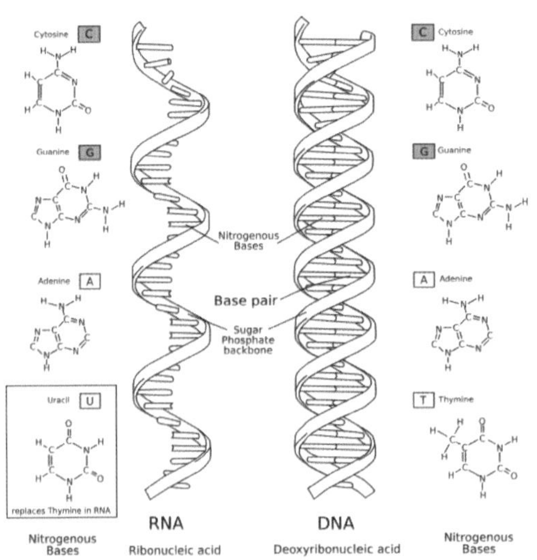

Left: An RNA strand, with its nitrogenous bases. Right: Double-stranded DNA.

Mutations involve a change in the base pair of an organism's genetic material. In most organisms, this means the mutation impacts the base pairs of deoxyribonucleic acid (DNA). In some cases, it may mean a change in the base pairs of ribonucleic acid (RNA).

DNA and RNA are nucleic acids. Nucleic acids are complex, high-molecular-weight macromolecule composed of polymers of repeating units (called monomers). Specifically, they consist of long chains of nucleotide monomers connected by covalent chemical bonds. A nucleotide is a chemical compound with three components: a nitrogen-containing base, a pentose (five-carbon) sugar, and one or more phosphate groups. The nitrogen-containing base of a nucleotide (also called the nucleobase) is typically a derivative of either purine or pyrimidine. The most common nucleotide bases are the purines adenine and guanine and the pyrimidines cytosine and thymine (or uracil in RNA). RNA molecules may contain as few as 75 nucleotides or more than 5,000 nucleotides, while a DNA molecule may be composed of more than 1,000,000 nucleotide units. The sugar component is either deoxyribose or ribose, giving the name of DNA and RNA. ("Deoxy" simply indicates that the sugar lacks an oxygen atom present in ribose, the parent compound).

The main role of DNA is the long-term storage of genetic information. DNA is often compared to a blueprint, since it contains instructions for constructing other components of the cell, such as proteins and RNA molecules. The DNA segments that carry genetic information are called genes, but other DNA sequences have structural purposes or are involved in regulating the expression of genetic information.

RNA serves as a genetic blueprint for certain viruses. However, it plays a diversity of roles in other organisms. RNA may be thought of as the intermediate between the DNA blueprint and the actual workings of the cell, serving as the template for the synthesis of proteins from the genetic information stored in DNA. Some RNA molecules (called ribozymes) are also involved in the catalysis of biochemical reactions.

In other words, chemically, DNA is a long polymer of simple units called nucleotides, with a backbone made of sugars (deoxyribose) and phosphate atoms joined by ester bonds. Attached to each sugar is one of four types of molecules called bases: adenine (A), guanine (G), cytosine (C), or thymine (T). Likewise, RNA is a nucleic acid consisting of chains of nucleotides also forming a polymer, with each nucleotide consisting of a nitrogenous base (adenine, cytosine, guanine, or uracil, U), ribose as the sugar component, and a phosphate.

It is the sequence of these four bases along the backbone of DNA that encodes information. This information is read using the genetic code, which specifies the sequence of the amino acids within proteins. It reads it in a triplet of bases, with one triplet translating into a particular amino acid. That is, the codon adenine-guanine-cytosine may translate to one particular amino acid, while the codon adenine-guanine-adenine may translate to another amino acid. The code is read by copying stretches of DNA into the related nucleic acid RNA, in a process called transcription. Most of these RNA molecules are used to synthesize proteins.

A mutation is a change in the sequence of the four bases along the backbone of DNA (or RNA). As a result, the sequence of amino acids may be changed, which would affect the structure of the protein that is encoded. However, not all changes in the bases (such as a mutation involving a change from adenine to guanine) necessarily results in the replacement with another amino acid, since there is code redundancy, with some different sequences of bases translating into the same amino acid.

The functionality of a protein is highly dependent on its three-dimensional structure (how it "folds") and this is highly dependent on the order of the amino acids that make it up. A change in a single amino acid could make the protein non-functional.

Within cells, DNA is organized into structures called chromosomes and the set of chromosomes within a cell make up a genome. These chromosomes are duplicated before cells divide, in a process called DNA replication. Eukaryotic organisms such as animals, plants, and fungi store their DNA inside the cell nucleus, while in prokaryotes such as bacteria it is found in the cell's cytoplasm. Within the chromosomes, chromatin proteins such as histones compact and organize DNA, which helps control its interactions with other proteins and thereby control which genes are transcribed.

Classifications of Mutations

By Effect on Structure

The sequence of a gene can be altered in a number of ways. Gene mutations have varying effects on health depending on where they occur and whether they alter the function of essential proteins. Structurally, mutations can be classified as:

- Small-scale mutations. These are mutations that affecting one or a few nucleotides. These include:
    - Point mutations. Point mutations involve an exchange of a single nucleotide for another one. Most common is the transition that exchanges a purine for a purine (A ↔ G) or a pyrimidine for a pyrimidine, (C ↔ T). Point mutations are often caused by chemicals or malfunction of DNA replication. A transition can be caused by nitrous acid, base mis-pairing, or mutagenic base analogs such as 5-bromo-2-deoxyuridine (BrdU). Less common is a transversion, which exchanges a purine for a pyrimidine or a pyrimidine for a purine (C/T ↔ A/G). A point mutation can be reversed by another point mutation, in which the nucleotide is changed back to its original state (true reversion) or by second-site reversion (a complementary mutation elsewhere that results in regained gene functionality). These changes are classified as transitions or transversions. An example of a transversion is adenine (A) being converted into a cytosine (C). There are also many other examples that can be found. Another Point mutation that occur within the protein coding region of a gene may be classified into three kinds, depending upon what the erroneous codon codes for:
- Silent mutations. Silent mutations are those that code for the same amino acid, because of the redundancy of the genetic code. That is, there is a change in the bases but the altered triplet still inserts the same amino acid as before into the protein.
- Missense mutations. Missense mutations code for a different amino acid.
- Nonsense mutations. Nonsense mutations code for a "stop" and can truncate the protein.
    - Insertions. Insertions are mutations that add one or more extra nucleotides into the DNA. They are usually caused by transposable elements, or errors during replication of repeating elements (e.g. AT repeats). Insertions in the coding region of a gene may alter splicing of the mRNA (splice site mutation), or cause a shift in the reading frame

(frameshift), both of which can significantly alter the gene product. Insertions can be reverted by excision of the transposable element.

- Deletions. Deletions remove one or more nucleotides from the DNA. Like insertions, these mutations can alter the reading frame of the gene. They are irreversible.

- Large-scale mutations. Large scale mutations in chromosomal structure may include:
  - Amplifications. Amplifications (or gene duplications) lead to multiple copies of all chromosomal regions, increasing the dosage of the genes located within them.
  - Deletions. Deletions of large chromosomal regions lead to loss of the genes within those regions.
  - Juxtaposing mutations. Mutations whose effect is to juxtapose previously separate pieces of DNA, potentially bringing together separate genes to form functionally distinct fusion genes (e.g. bcr-abl). These include:
    1. Chromosomal translocations. Chromosomal translocations interchange genetic parts from nonhomologous chromosomes.
    2. Interstitial deletions. Intersitional deletions involve an intra-chromosomal deletion that removes a segment of DNA from a single chromosome, thereby apposing previously distant genes. For example, cells isolated from a human astrocytoma, a type of brain tumor, were found to have a chromosomal deletion removing sequences between the "fused in glioblastoma" (fig) gene and the receptor tyrosine kinase "ros," producing a fusion protein (FIG-ROS). The abnormal FIG-ROS fusion protein has constitutively active kinase activity that causes oncogenic transformation (a transformation from normal cells to cancer cells).
    3. Chromosomal inversions. Chromosomal inversions involve reversing the orientation of a chromosomal segment.
  - Loss of heterozygosity. Loss of heterozygosity involves the loss of one allele, either by a deletion or recombination event, in an organism that previously had two different alleles.

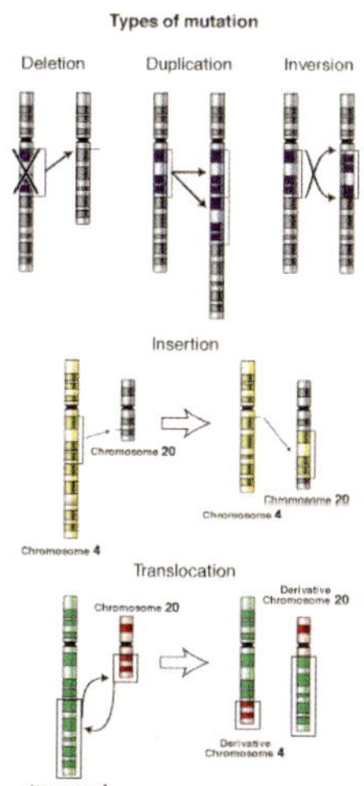

Illustrations of five types of chromosomal mutations.

By Effect on Function

- Loss-of-function mutations: Loss-of-function mutations are the result of gene product having less or no function. When the allele has a complete loss of function (null allele) it is often called an amorphic mutation. Phenotypes associated with such mutations are most often recessive. Exceptions are when the organism is haploid, or when the reduced dosage of a normal gene product is not enough for a normal phenotype (this is called haploinsufficiency).
- Gain-of-function mutations: Gain-of-function mutations change the gene product such that it gains a new and abnormal function. These mutations usually have dominant phenotypes. Often called a neo-morphic mutation.
- Dominant negative mutations: Dominant negative mutations (also called anti-morphic mutations) have an altered gene product that acts antagonistically to the wild-type allele. These mutations usually result in an altered molecular function (often inactive) and are characterized by a dominant or semi-dominant phenotype. In humans, Marfan syndrome is an example of a dominant negative mutation occurring in an autosomal dominant disease. In this condition, the defective glycoprotein product of the fibrillin gene (FBN1) antagonizes the product of the normal allele.
- Lethal mutations: Lethal mutations are mutations that lead to a phenotype incapable of effective reproduction.

By Aspect of Phenotype Affected

- Morphological mutations: Morphological mutations usually affect the outward appearance of an individual. Mutations can change the height of a plant or change it from smooth to rough seeds.
- Biochemical mutations: Biochemical mutations result in lesions stopping the enzymatic pathway. Often, morphological mutants are the direct result of a mutation due to the enzymatic pathway.

By Inheritance

The human genome contains two copies of each gene—a paternal and a maternal allele:

- Wildtype or Homozygous non-mutated: This occurs when neither of the alleles are mutated.
- A Heterozygous mutation: Heterozygous mutations occurs when only one allele is mutated.
- A Homozygous mutation: Homozygous mutation is when both the paternal and maternal alleles have an identical mutation.
- Compound heterozygous mutations: Compound heterozygous mutations or a genetic compound is when the paternal and maternal alleles have two different mutations.

Special Classes

- Conditional mutation: Conditional mutation is a mutation that has wild-type (or less severe) phenotype under certain "permissive" environmental conditions and a mutant phenotype under certain "restrictive" conditions. For example, a temperature-sensitive mutation can cause cell death at high temperature (restrictive condition), but might have no deleterious consequences at a lower temperature (permissive condition).

Causes of Mutation

Two classes of mutations are spontaneous mutations (molecular decay) and induced mutations caused by mutagens.

Spontaneous mutations- Spontaneous mutations on the molecular level include:

- Tautomerism: Tautomerism involves a base being changed by the repositioning of a hydrogen atom.
- Depurination: Depurination involves the loss of a purine base (A or G).
- Deamination: Deamination is a change in a normal base to an atypical base; C → U, (which can be corrected by DNA repair mechanisms), or spontaneous deamination of 5-methycytosine (irreparable), or A → HX (hypoxanthine).
- Transition: Transition occurs when a purine changes to another purine, or a pyrimidine to a pyrimidine.
- Transversion: Transversion occurs when a purine becomes a pyrimidine, or vice versa.

Induced mutations. Induced mutations on the molecular level can be caused by:

- Chemicals
    - Nitrosoguanidine (NTG)
    - Hydroxylamine NH_2OH
    - Base analogs (e.g., BrdU)
    - Simple chemicals (e.g., acids)
    - Alkylating agents (e.g., N-ethyl-N-nitrosourea (ENU)). These agents can mutate both replicating and non-replicating DNA. In contrast, a base analog can only mutate the DNA when the analog is incorporated in replicating the DNA. Each of these classes of chemical mutagens has certain effects that then lead to transitions, transversions, or deletions.
    - Methylating agents (e.g., ethyl methanesulfonate (EMS))
    - Polycyclic hydrocarbons (e.g., benzopyrenes found in internal combustion engine exhaust)
    - DNA intercalating agents (e.g., ethidium bromide)
    - DNA cross linker (e.g., platinum)
    - Oxidative damage, caused by oxygen(O)] radicals.

- Radiation
    - Ultraviolet: Ultraviolet radiation (nonionizing radiation) excites electrons to a higher energy level. DNA absorbs one form, ultraviolet light. Two nucleotide bases in DNA, cytosine and thymine, are most vulnerable to excitation that can change base-pairing properties. UV light can induce adjacent thymine bases in a DNA strand to pair with each other, as a bulky dimer.
    - Ionizing radiation

DNA has so-called hotspots, where mutations occur up to 100 times more frequently than the normal mutation rate. A hotspot can be at an unusual base, e.g., 5-methylcytosine.

Mutation rates also vary across species. Evolutionary biologists have theorized that higher mutation rates are beneficial in some situations, because they allow organisms to evolve and therefore adapt more quickly to their environments. For example, repeated exposure of bacteria to antibiotics, and selection of resistant mutants, can result in the selection of bacteria that have a much higher mutation rate than the original population (mutator strains).

Nomenclature

Nomenclature of mutations specify the type of mutation and base or amino acid changes.

- Amino acid substitution: For example, D111E. The first letter is the one letter code of the wildtype amino acid, the number is the position of the amino acid from the N terminus, and the second letter is the one letter code of the amino acid present in the mutation. If the second letter is 'X', any amino acid may replace the wildtype.
- Amino acid deletion: For example, ΔF508. The Greek symbol Δ or 'delta' indicates a deletion. The letter refers to the amino acid present in the wildtype and the number is the position from the N terminus of the amino acid were it to be present as in the wildtype.

Types of Mutations

Adaptive Mutation

In mainstream biological thought, it is held that while mutagenesis is non-random in many ways, the utility of a genetic mutation to the organism in which it occurs does not affect the rate at which it occurs. However experimental evidence exists that in some instances the rate of specific mutations arising is greater when they are advantageous to the organism than when they are not.

Back Mutation

Back mutation is a change in a nucleotide pair of a point-mutated DNA sequence that restores the original sequence and hence the original phenotype.

Frameshift Mutation

A frameshift mutation is a mutation caused by indels, i.e.. inserts or deletes in a number of nucleotides that is not evenly divisible by three from a DNA sequence. Due to the triplet nature of gene expression by codons, the insertion or deletion can disrupt the reading frame, or the grouping of the codons, resulting in a completely different translation from the original. The earlier in the sequence the deletion or insertion occurs, the more altered the protein produced is.

Missense Mutation

Missense mutations or nonsynonymous mutations are types of point mutations where a single nucleotide is changed to cause substitution of a different amino acid. This in turn can render the resulting protein nonfunctional. Such mutations are responsible for diseases such as Epidermolysis bullosa, sickle-cell disease, and SOD1 mediated ALS.

Neutral Mutation

A neutral mutation is a mutation that occurs in an amino acid codon (presumably within an mRNA molecule) that results in the substitution of a different, but chemically similar, amino acid. This is similar to a silent mutation, where a codon mutation may encode the same amino acid; for example, a change from AUU to AUC will still encode leucine, so no discernable change occurs (a silent mutation).

Nonsense Mutation

A nonsense mutation is a point mutation in a sequence of DNA that results in a premature stop codon, or a nonsense codon in the transcribed mRNA, and possibly a truncated, and often nonfunctional protein product.

Point Mutation

A point mutation, or substitution, is a type of mutation that causes the replacement of a single base nucleotide with another nucleotide. Often the term point mutation also includes insertions or deletions of a single base pair (which have more of an adverse effect on the synthesized protein due to nucleotides still being read in triplets, but in different frames: a mutation called a frameshift mutation).

Silent Mutation

Silent mutations are DNA mutations that do not result in a change to the amino acid sequence of a protein. They may occur in a non-coding region (outside of a gene or within an intron), or they may occur within an exon in a manner that does not alter the final amino acid sequence. The phrase silent mutation is often used interchangeably with the phrase synonymous mutation; however, synonymous mutations are a subcategory of the former, occurring only within exons.

Harmful Mutations

Changes in DNA caused by mutation can cause errors in protein sequence, creating partially or completely non-functional proteins. To function correctly, each cell depends on thousands of proteins to function in the right places at the right times. When a mutation alters a protein that plays a critical role in the body, a medical condition can result. A condition caused by mutations in one or more genes is called a genetic disorder. However, only a small percentage of mutations cause genetic disorders; most have no impact on health. For example, some mutations alter a gene's DNA base sequence but do not change the function of the protein made by the gene.

If a mutation is present in a germ cell, it can give rise to offspring that carries the mutation in all of its cells. This is the case in hereditary diseases. On the other hand, a mutation can occur in a somatic cell of an organism. Such mutations will be present in all descendants of this cell, and certain mutations can cause the cell to become malignant, and thus cause cancer.

Often, gene mutations that could cause a genetic disorder are repaired by the DNA repair system of the cell. Each cell has a number of pathways through which enzymes recognize and repair mistakes in DNA. Because DNA can be damaged or mutated in many ways, the process of DNA repair is an important way in which the body protects itself from disease.

Beneficial Mutations

A very small percentage of all mutations actually have a positive effect. These mutations lead to new versions of proteins that help an organism and its future generations better adapt to changes in their environment.

For example, a specific 32 base pair deletion in human CCR5 (CCR5-Δ32) confers HIV resistance to homozygotes and delays AIDS onset in heterozygotes. The CCR5 mutation is more common in those of European descent. One theory for the etiology of the relatively high frequency of CCR5-Δ32 in the European population is that it conferred resistance to the bubonic plague in mid-fourteenth century Europe. People who had this mutation were able to survive infection thus its frequency in the population increased. It could also explain why this mutation is not found in Africa where the bubonic plague never reached. A more recent theory says the selective pressure on the CCR5 Delta 32 mutation has been caused by smallpox instead of bubonic plague.

Darwin's Theory

Carles Darwin is known as the father of evolution due to his contribution to the establishment of the theory of evolution. His theory helped in removing all the conventional old believes which said that the formation of various species was a supernatural phenomenon or act of the Almighty. Darwin's evolutionary theory of natural selection gave a more rational explanation of the formation of new species. As per natural selection, various species originated from a single species as a result of adaptation to the changing environment.

Darwin and the Voyage of the Beagle

Darwin's seminal book, On the Origin of Species, set forth his ideas about evolution and natural selection. These ideas were largely based on direct observations from Darwin's travels around the globe. From 1831 to 1836, he was part of a survey expedition carried out by the ship HMS Beagle, which included stops in South America, Australia, and the southern tip of Africa. At each of the expedition's stops, Darwin had the opportunity to study and catalog the local plants and animals.

Over the course of his travels, Darwin began to see intriguing patterns in the distribution and features of organisms. We can see some of the most important patterns Darwin noticed in distribution of organisms by looking at his observations of the Galápagos Islands off the coast of Ecuador.

Darwin found that nearby islands in the Galápagos had similar but nonidentical species of finches living on them. Moreover, he noted that each finch species was well-suited for its environment and role. For instance, species that ate large seeds tended to have large, tough beaks, while those that ate insects had thin, sharp beaks. Finally, he observed that the finches (and other animals) found on the Galápagos Islands were similar to species on the nearby mainland of Ecuador, but different from those found elsewhere in the world.

Darwin didn't figure all of this out on his trip. In fact, he didn't even realize all the finches were related but distinct species until he showed his specimens to a skilled ornithologist (bird biologist) years later. Gradually, however, he came up with an idea that could explain the pattern of related but different finches.

According to Darwin's idea, this pattern would make sense if the Galápagos Islands had long ago been populated by birds from the neighboring mainland. On each island, the finches might have gradually adapted to local conditions (over many generations and long periods of time). This process could have led to the formation of one or more distinct species on each island.

If this idea was correct, though, why was it correct? What mechanism could explain how each finch population had acquired adaptations, or features that made it well-suited to its immediate environment? During his voyage, and in the years after, Darwin developed and refined a set of ideas that could explain the patterns he had observed during his voyage. In his book, On the Origin of Species, Darwin outlined his two key ideas: evolution and natural selection.

"Darwin's finches," by John Gould

Evolution

Darwin proposed that species can change over time, that new species come from pre-existing species, and that all species share a common ancestor. In this model, each species has its own unique set of heritable (genetic) differences from the common ancestor, which have accumulated gradually over very long time periods. Repeated branching events, in which new species split off from a common ancestor, produce a multi-level "tree" that links all living organisms.

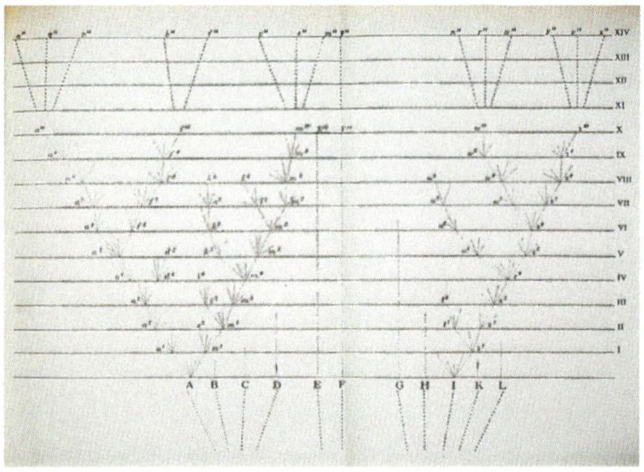

Modern-day species appear at the top of the chart, while the ancestors from which they arose are shown lower in the chart.

Darwin referred to this process, in which groups of organisms change in their heritable traits over generations, as "descent with modification." Today, we call it evolution. Darwin's sketch above illustrates his idea, showing how one species can branch into two over time, and how this process can repeat multiple times in the "family tree" of a group of related species.

Natural Selection

Importantly, Darwin didn't just propose that organisms evolved. If that had been the beginning and end of his theory, he wouldn't be in as many textbooks as he is today! Instead, Darwin also proposed a mechanism for evolution: natural selection. This mechanism was elegant and logical, and it explained how populations could evolve (undergo descent with modification) in such a way that they became better suited to their environments over time.

Darwin's concept of natural selection was based on several key observations:

- Traits are often heritable: In living organisms, many characteristics are inherited, or passed from parent to offspring. (Darwin knew this was the case, even though he did not know that traits were inherited via genes).

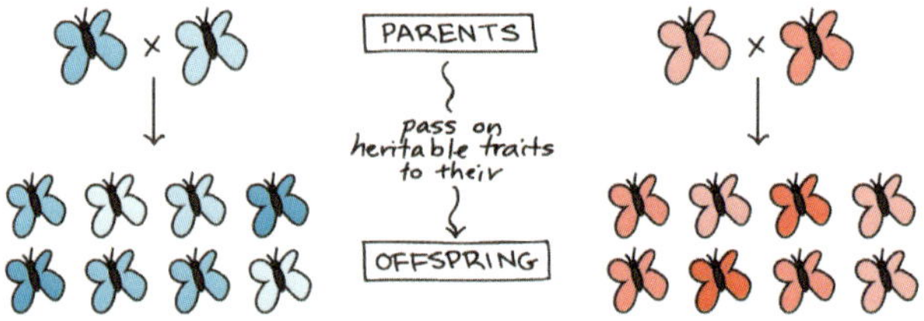

- More offspring are produced than can survive: Organisms are capable of producing more offspring than their environments can support. Thus, there is competition for limited resources in each generation.

- Offspring vary in their heritable traits. The offspring in any generation will be slightly different from one another in their traits (color, size, shape, etc.), and many of these features will be heritable.

Based on these simple observations, Darwin concluded the following:

- In a population, some individuals will have inherited traits that help them survive and reproduce (given the conditions of the environment, such as the predators and food sources present). The individuals with the helpful traits will leave more offspring in the next generation than their peers, since the traits make them more effective at surviving and reproducing.
- Because the helpful traits are heritable, and because organisms with these traits leave more offspring, the traits will tend to become more common (present in a larger fraction of the population) in the next generation.
- Over generations, the population will become adapted to its environment (as individuals with traits helpful in that environment have consistently greater reproductive success than their peers).

Darwin's model of evolution by natural selection allowed him to explain the patterns he had seen during his travels. For instance, if the Galápagos finch species shared a common ancestor, it made sense that they should broadly resemble one another (and mainland finches, who likely shared that common ancestor). If groups of finches had been isolated on separate islands for many generations, however, each group would have been exposed to a different environment in which different heritable traits might have been favored, such as different sizes and shapes of beaks for using different food sources. These factors could have led to the formation of distinct species on each island.

Working of Natural Selection

To make natural selection more concrete, let's consider a simplified, hypothetical example. In this example, a group of mice with heritable variation in fur color (black vs. tan) has just moved into a new area where the rocks are black. This environment features hawks, which like to eat mice and can see the tan ones more easily than the black ones against the black rock.

Because the hawks can see and catch the tan mice more easily, a relatively large fraction of the tan mice are eaten, while a much smaller fraction of the black mice are eaten. If we look at the ratio of black mice to tan mice in the surviving ("not-eaten") group, it will be higher than in the starting population.

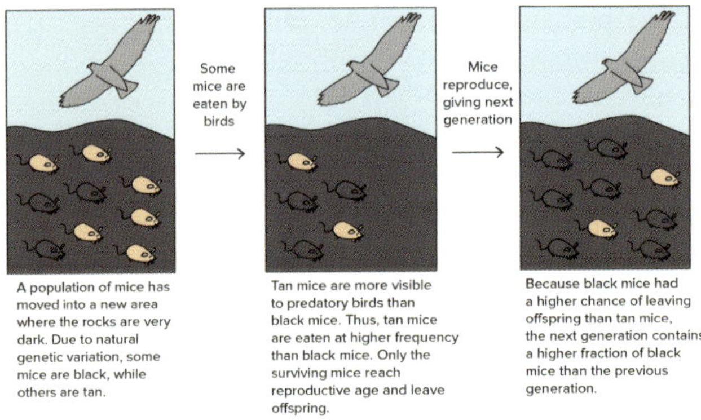

Black and white line art drawing of Swainson hawk bird in flight

Fur color is a heritable trait (one that can be passed from parent to child). So, the increased fraction of black mice in the surviving group means an increased fraction of black baby mice in the next generation. After several generations of selection, the population might be made up almost entirely of black mice. This change in the heritable features of the population is an example of evolution.

Natural Selection Depends on the Environment

Natural selection doesn't favor traits that are somehow inherently superior. Instead, it favors traits that are beneficial (that is, help an organism survive and reproduce more effectively than its peers) in a specific environment. Traits that are helpful in one environment might actually be harmful in another.

For example, in the simplified scenario above, the black mice don't become more common over generations because they are inherently "better" or "more evolved" than tan mice. Instead, they become more common because they have a heritable feature that makes them better able to survive and reproduce in a specific setting, one that happens to include black rocks. In a setting with light-colored rocks, the helpful and harmful traits would be reversed.

Natural Selection Acts on Riation

Natural selection needs some starting material, and that starting material is heritable variation. For natural selection to act on a feature, there must already be variation (differences among individuals) for that feature. Also, the differences have to be heritable, determined by the organisms' genes.

For instance, in the mouse example, there was heritable variation in fur color: certain gene variants made a mouse black, while others made it tan. If all the mice had been tan, the population would have had no way to adapt to its new environment by natural selection and might instead have been wiped out.

What if there had been non-heritable variation (e.g., because we dyed some tan mice black)? This might have affected the survival of the individuals, but it wouldn't have changed the composition of the population over generations (because the babies of the dyed tan mice would have inherited the tan-mouse gene variants present in their parents, not black-mouse ones).

Heritable Variation Comes From Random Mutations

The original source of the new gene variants that produce new heritable traits, such as fur colors, is random mutation (changes in DNA sequence). Random mutations that are passed on to offspring typically occur in the germline, or sperm and egg cell lineage, of organisms. Sexual reproduction "mixes and matches" gene variants to make more variation.

References

- What-is-evolution: yourgenome.org, Retrieved 25 May 2018
- What-is-genetic-variation: yourgenome.org, Retrieved 11 April 2018
- Diversity-of-Life-227556259: researchgate.net, Retrieved 31 March 2018
- Mutation: newworldencyclopedia.org, Retrieved 15 May 2018
- Darwins-contribution-theory-evolution: byjus.com, Retrieved 31 June 2018

5

Reproduction

Reproduction is a natural process that occurs in living organisms by which new individuals are created. Reproduction is of two types, sexual and asexual. The topics elaborated in this chapter on plant reproduction, animal reproduction and human reproduction system will help in providing a comprehensive understanding of the processes of reproduction.

Production of new individuals along a leaf margin of the air plant, Kalanchoë pinnata. The small plant in front is about 1 cm tall. The concept of "individual" is obviously stretched by this asexual reproductive process.

In biology, reproduction is the process by which new individual organisms are produced. Reproduction is a fundamental feature of all known life; each individual organism exists as the result of reproduction. Although the term reproduction encompasses a great variety of means by which organisms produce new offspring, reproductive processes can be classified into two main types: Sexual reproduction and asexual reproduction.

Sexual reproduction is a biological process by which organisms create descendants that have a combination of genetic material contributed from two different gametes (reproductive or sex cells), usually from two different organisms. The union (fusion) of these gametes produces an organism that is not genetically identical to the parent(s). Sexual reproduction generally involves contribution of genetic material to the offspring from two different organisms, but includes self-fertilization within one organism but involving fusion of two gametes. Humans are an example of a species that reproduces by sexual reproduction involving two individuals. Peanuts are a type of plant that can self-pollinate (self-fertilize).

In asexual reproduction, an individual can reproduce without involvement with another individual, there is no fusion of gametes, and the new organism produced inherits all of its chromosomes from one parent and thus is a genetically-similar or identical copy of the parent. The division of a bacterial cell into two daughter cells is an example of asexual reproduction. Asexual reproduction is not, however, limited to single-celled organisms. Most plants have the ability to reproduce asexually.

In general, more-complex organisms reproduce sexually while simpler, usually unicellular, organisms reproduce asexually. Among animals, and indeed eukaryotes in general, sexual reproduction is a nearly universal mode of reproduction. However, even lower plants, fungi, some of the protists, and bacteria likewise exhibit reproductive variances, which may be symbolized by + and - signs (rather than being called male and female), and referred to as "mating strains" or "reproductive types" or similar appellations. This polarity reflects the principle of dual characteristics, whereby existent beings exhibit the quality of positivity and negativity.

While reproductive processes can be classified into two broad groupings, species exhibit a wide spectrum of mechanisms. For example, some plants alternate between sexual and asexual reproduction. In parthenogenesis, such as found in some invertebrates (rotifers, water fleas, aphids, some bees, etc.) and some vertebrates (some amphibians, reptiles, and more rarely in birds), an embryo is produced without fertilization by a male. Generally, authors list parthenogenesis as a form of asexual reproduction because it does not involve fusion of gametes of opposite sexes, nor any exchange of genetic material from two different sources; however, some authorities (McGraw-Hill 2004) classify parthenogenesis as sexual reproduction on the basis that it involves gametes or does not produce an offspring genetically identical to the parent.

Asexual vs. Sexual Reproduction

Sexual reproduction is a near-universal mode of reproduction among eukaryotes. Furthermore, while asexual reproduction is widespread among fungi, certain protists and vascular plants, various invertebrates, and even some reptiles and amphibians, sexual reproduction is also seen in these same groups, and some state even in bacteria, which will exchange genetic material between donors (+ mating type) and recipients (- mating type). (However, many authorities, such as Mayr, consider the unidirectional lateral transfer of genetic material in bacteria to either not be reproduction, or at least not sexual reproduction, and Mayr states that sexual reproduction is unknown in prokaryotes.) Lahr contend that even in amoeboid lineages the extent of asexual reproduction in overestimated and that the evidence "demonstrates that the majority of amoeboid lineages are, contrary to popular belief, anciently sexual, and that most asexual groups have probably arisen recently and independently." Among animals, nearly all species practice sexual reproduction. Mayr notes that "above the level of the genus there are only three higher taxa of animals that consist exclusively of uniparentally reproducing clones." (An example of these higher taxa would be rotifers of the Class Bdelloidea).

Why sexual reproduction appeared and is so prevalent is a major puzzle in modern biology. Sexual reproduction has many drawbacks, since it requires far more energy than asexual reproduction. For example, in an asexual species, each member of the population is capable of bearing young, implying that an asexual population can grow more rapidly. An additional cost of sexual reproduction is that males and females must search for each other in order to mate. Evolutionary biologist and geneticist John Maynard Smith maintains that the perceived advantage for an individual organism to pass only its own entire genome to its offspring is so great that there must an advantage by at least a factor of two to explain why nearly all animal species maintain a male sex. Mayr notes that since the 1880s evolutionists have argued over the advantage of sexual reproduction and "so far, no clear-cut winner has emerged from this controversy."

General explanations for the origin and maintenance of sexual reproduction focus on the advantages conferred due to an improvement in the quality of progeny (fitness), despite reducing the overall number of offspring (two-fold cost of sex). This enhanced fitness is explained in terms of the genetic variation that is increased through sexual reproduction. Organisms that reproduce through asexual reproduction tend to grow in number exponentially. However, because they rely on mutations for variations in their DNA, all members of the species have similar vulnerabilities. Organisms that reproduce sexually yield a smaller amount of offspring, but the large amount of variation in their genes makes them less susceptible to disease or changing environmental stresses.

For example, many organisms can reproduce sexually as well as asexually. Aphids, slime molds, sea anemones, some species of starfish (by fragmentation), and many plants are examples. It is held that when environmental factors are favorable, asexual reproduction is employed to exploit suitable conditions for survival, such as an abundant food supply, adequate shelter, favorable climate, disease, optimum pH, or a proper mix of other lifestyle requirements. Populations of these organisms increase exponentially via asexual reproductive strategies to take full advantage of the rich supply resources. When food sources have been depleted, the climate becomes hostile, or individual survival is jeopardized by some other adverse change in living conditions, it is held that these organisms switch to sexual forms of reproduction. The variations found in offspring of sexual reproduction allow some individuals to be better suited for survival and provide a mechanism for selective adaptation to occur. In addition, sexual reproduction usually results in the formation of a life stage that is able to endure the conditions that threaten the offspring of an asexual parent. Thus, seeds, spores, eggs, pupae, cysts or other "over-wintering" stages of sexual reproduction ensure the survival during unfavorable times and the organism can "wait out" adverse situations until a swing back to suitability occurs.

George C. Williams introduced the *lottery principle* in 1975 to explain this basic concept, using lottery tickets as an analogy. He argued that asexual reproduction, which produces little or no genetic variety in offspring, was like buying a large number of tickets that all have the same number, limiting the chance of "winning"—that is, surviving. Sexual reproduction, he argued, was like purchasing fewer tickets but with a greater variety of numbers and therefore a greater chance of success. The point of this analogy is that since asexual reproduction does not produce genetic variations, there is little ability to quickly adapt to a changing environment. The lottery principle is less accepted these days because of evidence that asexual reproduction is more prevalent in unstable environments, the opposite of what it predicts.

Conversely, Heng proposes that the resolution to the "paradox of sex" is that sexual reproduction actually reduces the drastic genetic diversity at the genome or chromosome level, resulting in the preservation of species identity, rather than the provision of evolutionary diversity for future environmental challenges. He maintains that while genetic recombination contributes to genetic diversity, it does so secondarily and within the framework of the chromosomally defined genome. That is, the asexual process generates more diverse genomes because of the less controlled reproduction systems, while sexual reproduction generates more stable genomes.

Reproductive Strategies

There is a wide range of reproductive strategies employed by different species. Some animals, such as Homo sapiens and Northern Gannet, do not reach sexual maturity for many years after birth and even then produce few offspring. Others reproduce quickly; but, under normal circumstances, most offspring do not survive to adulthood. For example, a rabbit (mature after 8 months) can produce 10–30 offspring per year, and a fruit fly (mature after 10–14 days) can produce up to 900 offspring per year.

These two main strategies are known as K-selection (few offspring) and r-selection (many offspring). Which strategy is favored depends on a variety of circumstances. Animals with few offspring can devote more resources to the nurturing and protection of each individual offspring, thus reducing the need for a large number of offspring. On the other hand, animals with many offspring may devote less resources to each individual offspring; for these types of animals it, is common for a large number of offspring to die soon after birth, but normally enough individuals survive to maintain the population.

Other types of reproductive strategies include polycyclic animals' (reproduce intermittently throughout their lives), Semelparous organisms (reproduce only once in their lifetime, such as annual plants, which often die shortly after reproduction), and Iteroparous organisms (produce offspring in successive cycles, such as perennial plants, and thus survive over multiple seasons).

Plant Reproduction

Reproduction in plants is either asexual or sexual. Asexual reproduction in plants involves a variety of widely disparate methods for producing new plants identical in every respect to the parent. Sexual reproduction, on the other hand, depends on a complex series of basic cellular events, involving chromosomes and their genes, that take place within an elaborate sexual apparatus evolved precisely for the development of new plants in some respects different from the two parents that played a role in their production.

Sexual Reproduction

There are several different methods and processes involved in the sexual reproduction of plants. Many of the structures associated with sexual reproduction in plants are valuable commodities for humans (think fruits, berries, and vegetables). An equal number of them are "baneful" (think seasonal allergies).

Self-pollination and Cross-pollination

In angiosperms, pollination is defined as the placement or transfer of pollen from the anther to the stigma of the same flower or another flower. In gymnosperms, pollination involves pollen transfer from the male cone to the female cone. Upon transfer, the pollen germinates to form the pollen tube and the sperm for fertilizing the egg. Pollination has been well studied since the time of Gregor Mendel. Mendel successfully carried out self- as well as cross-pollination in garden peas while studying how characteristics were passed on from one generation to the next. Today's crops are a result of plant breeding, which employs artificial selection to produce the present-day cultivars. A case in point is today's corn, which is a result of years of breeding that started with its ancestor, teosinte. The teosinte that the ancient Mayans originally began cultivating had tiny seeds—vastly different from today's relatively giant ears of corn. Interestingly, though these two plants appear to be entirely different, the genetic difference between them is miniscule.

Pollination takes two forms: self-pollination and cross-pollination. Self-pollination occurs when the pollen from the anther is deposited on the stigma of the same flower, or another flower on the same plant. Cross-pollination is the transfer of pollen from the anther of one flower to the stigma of another flower on a different individual of the same species. Self-pollination occurs in flowers where the stamen and carpel mature at the same time, and are positioned so that the pollen can land on the flower's stigma. This method of pollination does not require an investment from the plant to provide nectar and pollen as food for pollinators.

Living species are designed to ensure survival of their progeny; those that fail become extinct. Genetic diversity is therefore required so that in changing environmental or stress conditions, some of the progeny can survive. Self-pollination leads to the production of plants with less genetic diversity, since genetic material from the same plant is used to form gametes, and eventually, the zygote. In contrast, cross-pollination—or out-crossing—leads to greater genetic diversity because the microgametophyte and megagametophyte are derived from different plants.

Because cross-pollination allows for more genetic diversity, plants have developed many ways to avoid self-pollination. In some species, the pollen and the ovary mature at different times. These flowers make self-pollination nearly impossible. By the time pollen matures and has been shed, the stigma of this flower is mature and can only be pollinated by pollen from another flower. Some flowers have developed physical features that prevent self-pollination. The primrose is one such flower. Primroses have evolved two flower types with differences in anther and stigma length: the pin-eyed flower has anthers positioned at the pollen tube's halfway point, and the thrum-eyed flower's stigma is likewise located at the halfway point. Insects easily cross-pollinate while seeking the nectar at the bottom of the pollen tube. This phenomenon is also known as heterostyly. Many plants, such as cucumber, have male and female flowers located on different parts of the plant, thus making self-pollination difficult. In yet other species, the male and female flowers are borne on different plants (dioecious). All of these are barriers to self-pollination; therefore, the plants depend on pollinators to transfer pollen. The majority of pollinators are biotic agents such as insects (like bees, flies, and butterflies), bats, birds, and other animals. Other plant species are pollinated by abiotic agents, such as wind and water.

Methods of Pollination

Pollination by Insects

Figure: Insects, such as bees, are important agents of pollination.

Bees are perhaps the most important pollinator of many garden plants and most commercial fruit trees (figure). The most common species of bees are bumblebees and honeybees. Since bees cannot see the color red, bee-pollinated flowers usually have shades of blue, yellow, or other colors. Bees collect energy-rich pollen or nectar for their survival and energy needs. They visit flowers that are open during the day, are brightly colored, have a strong aroma or scent, and have a tubular shape, typically with the presence of a nectar guide. A nectar guide includes regions on the flower petals that are visible only to bees, and not to humans; it helps to guide bees to the center of the flower, thus making the pollination process more efficient. The pollen sticks to the bees' fuzzy hair, and when the bee visits another flower, some of the pollen is transferred to the second flower. Recently, there have been many reports about the declining population of honeybees. Many flowers will remain unpollinated and not bear seed if honeybees disappear. The impact on commercial fruit growers could be devastating.

Many flies are attracted to flowers that have a decaying smell or an odor of rotting flesh. These flowers, which produce nectar, usually have dull colors, such as brown or purple. They are found on the corpse flower or voodoo lily (Amorphophallus), dragon arum (Dracunculus), and carrion flower (Stapleia, Rafflesia). The nectar provides energy, whereas the pollen provides protein. Wasps are also important insect pollinators, and pollinate many species of figs.

Butterflies, such as the monarch, pollinate many garden flowers and wildflowers, which usually occur in clusters. These flowers are brightly colored, have a strong fragrance, are open during the day, and have nectar guides to make access to nectar easier. The pollen is picked up and carried on the butterfly's limbs. Moths, on the other hand, pollinate flowers during the late afternoon and night. The flowers pollinated by moths are pale or white and are flat, enabling the moths to land. One well-studied example of a moth-pollinated plant is the yucca plant, which is pollinated by the yucca moth. The shape of the flower and moth have adapted in such a way as to allow successful pollination. The moth deposits pollen on the sticky stigma for fertilization to occur later. The female moth also deposits eggs into the ovary. As the eggs develop into larvae, they obtain food from the flower and developing seeds. Thus, both the insect and flower benefit from each other in this symbiotic relationship. The corn earworm moth and Gaura plant have a similar relationship (figure).

Figure: A corn earworm sips nectar from a night-blooming Gaura plant.

Pollination by Bats

In the tropics and deserts, bats are often the pollinators of nocturnal flowers such as agave, guava, and morning glory. The flowers are usually large and white or pale-colored; thus, they can be distinguished from the dark surroundings at night. The flowers have a strong, fruity, or musky fragrance and produce large amounts of nectar. They are naturally large and wide-mouthed to accommodate the head of the bat. As the bats seek the nectar, their faces and heads become covered with pollen, which is then transferred to the next flower.

Pollination by Birds

Many species of small birds, such as the hummingbird (figure) and sun birds, are pollinators for plants such as orchids and other wildflowers. Flowers visited by birds are usually sturdy and are oriented in such a way as to allow the birds to stay near the flower without getting their wings entangled in the nearby flowers. The flower typically has a curved, tubular shape, which allows access for the bird's beak. Brightly colored, odorless flowers that are open during the day are pollinated by birds. As a bird seeks energy-rich nectar, pollen is deposited on the bird's head and neck and is then transferred to the next flower it visits. Botanists have been known to determine the range of extinct plants by collecting and identifying pollen from 200-year-old bird specimens from the same site.

Figure: Humming birds have adaptations that allow them to reach the nectar of certain tubular flowers.

Pollination by Wind

Most species of conifers, and many angiosperms, such as grasses, maples and oaks, are pollinated by wind. Pine cones are brown and unscented, while the flowers of wind-pollinated angiosperm species are usually green, small, may have small or no petals, and produce large amounts of pollen. Unlike the typical insect-pollinated flowers, flowers adapted to pollination by wind do not produce nectar or scent. In wind-pollinated species, the microsporangia hang out of the flower, and, as the wind blows, the lightweight pollen is carried with it (figure).

Figure: A person knocks pollen from a pine tree.

The flowers usually emerge early in the spring, before the leaves, so that the leaves do not block the movement of the wind. The pollen is deposited on the exposed feathery stigma of the flower (figure).

Figure: These male (a) and female (b) catkins are from the goat willow tree (Salix caprea).

Pollination by Water

Some weeds, such as Australian sea grass and pond weeds, are pollinated by water. The pollen floats on water, and when it comes into contact with the flower, it is deposited inside the flower.

Pollination by Deception

Orchids are highly valued flowers, with many rare varieties (figure). They grow in a range of specific habitats, mainly in the tropics of Asia, South America, and Central America. At least 25,000 species of orchids have been identified.

Flowers often attract pollinators with food rewards, in the form of nectar. However, some species of orchid are an exception to this standard: they have evolved different ways to attract the desired pollinators. They use a method known as food deception, in which bright colors and perfumes are offered, but no food. Anacamptis morio, commonly known as the green-winged orchid, bears bright purple flowers and emits a strong scent. The bumblebee, its main pollinator, is attracted to the flower because of the strong scent—which usually indicates food for a bee—and in the process, picks up the pollen to be transported to another flower.

Figure: Certain orchids use food deception or sexual deception to attract pollinators. Shown here is a bee orchid (Ophrys apifera).

Other orchids use sexual deception. Chiloglottis trapeziformis emits a compound that smells the same as the pheromone emitted by a female wasp to attract male wasps. The male wasp is attracted to the scent, lands on the orchid flower, and in the process, transfers pollen. Some orchids, like the Australian hammer orchid, use scent as well as visual trickery in yet another sexual deception strategy to attract wasps. The flower of this orchid mimics the appearance of a female wasp and emits a pheromone. The male wasp tries to mate with what appears to be a female wasp, and in the process, picks up pollen, which it then transfers to the next counterfeit mate.

Double Fertilization

After pollen is deposited on the stigma, it must germinate and grow through the style to reach the ovule. The microspores, or the pollen, contain two cells: the pollen tube cell and the generative cell. The pollen tube cell grows into a pollen tube through which the generative cell travels. The germination of the pollen tube requires water, oxygen, and certain chemical signals. As it travels through the style to reach the embryo sac, the pollen tube's growth is supported by the tissues of the style. In the meantime, if the generative cell has not already split into two cells, it now divides to form two sperm cells. The pollen tube is guided by the chemicals secreted by the synergids present in the embryo sac, and it enters the ovule sac through the micropyle. Of the two sperm cells, one sperm fertilizes the egg cell, forming a diploid zygote; the other sperm fuses with the two polar nuclei, forming a triploid cell that develops into the endosperm. Together, these two fertilization events in angiosperms are known as double fertilization (figure). After fertilization is complete, no other sperm can enter. The fertilized ovule forms the seed, whereas the tissues of the ovary become the fruit, usually enveloping the seed.

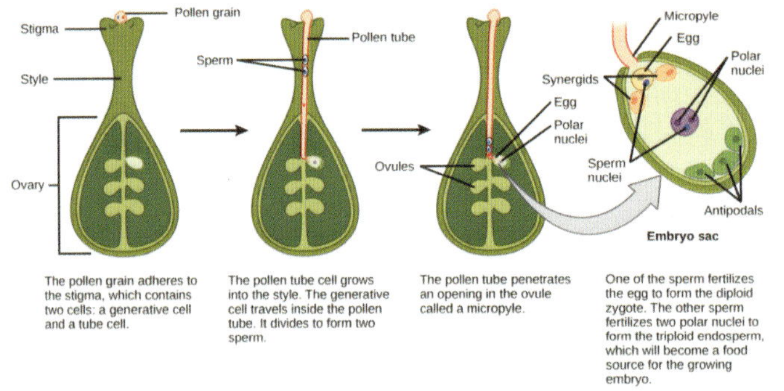

Figure: In angiosperms, one sperm fertilizes the egg to form the 2n zygote, and the other sperm fertilizes the central cell to form the 3n endosperm. This is called a double fertilization.

After fertilization, the zygote divides to form two cells: the upper cell, or terminal cell, and the lower, or basal, cell. The division of the basal cell gives rise to the suspensor, which eventually makes connection with the maternal tissue. The suspensor provides a route for nutrition to be transported from the mother plant to the growing embryo. The terminal cell also divides, giving rise to a globular-shaped proembryo (figure (a)). In dicots (eudicots), the developing embryo has a heart shape, due to the presence of the two rudimentary cotyledons (figure (b)). In non-endospermic dicots, such as Capsella bursa, the endosperm develops initially, but is then digested, and the food reserves are moved into the two cotyledons. As the embryo and cotyledons enlarge, they run out of room inside the developing seed, and are forced to bend (figure (c)). Ultimately, the embryo and cotyledons fill the seed (figure (d)), and the seed is ready for dispersal. Embryonic development is suspended after some time, and growth is resumed only when the seed germinates. The developing seedling will rely on the food reserves stored in the cotyledons until the first set of leaves begin photosynthesis.

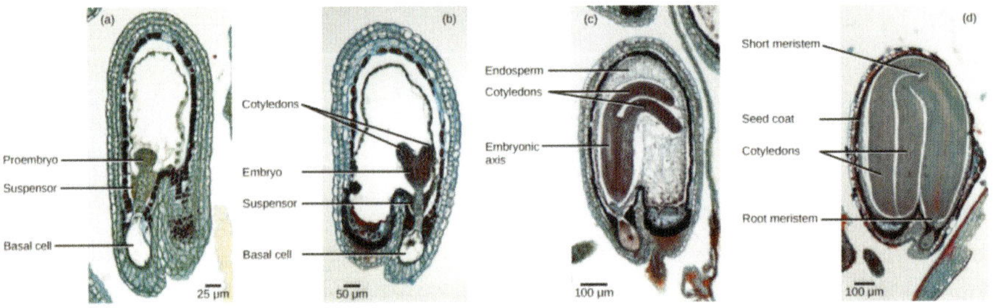

The stages of embryo development in the ovule of a shepherd's purse (Capsella bursa). After fertilization, the zygote divides to form an upper terminal cell and a lower basal cell. (a) In the first stage of development, the terminal cell divides, forming a globular pro-embryo. The basal cell also divides, giving rise to the suspensor. (b) In the second stage, the developing embryo has a heart shape due to the presence of cotyledons. (c) In the third stage, the growing embryo runs out of room and starts to bend. (d) Eventually, it completely fills the seed.

Development of a Seed

The mature ovule develops into the seed. A typical seed contains a seed coat, cotyledons, endosperm, and a single embryo (figure).

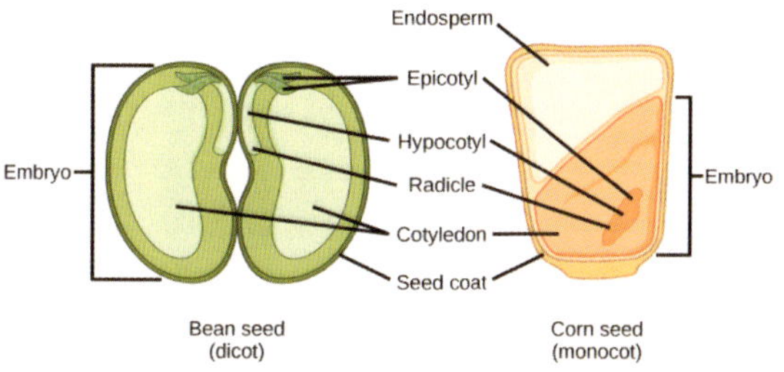

The structures of dicot and monocot seeds are shown. Dicots (left) have two cotyledons. Monocots, such as corn (right), have one cotyledon, called the scutellum; it channels nutrition to the growing embryo. Both monocot and dicot embryos have a plumule that forms the leaves, a hypocotyl that forms the stem, and a radicle that forms the root. The embryonic axis comprises everything between the plumule and the radicle, not including the cotyledon(s).

The storage of food reserves in angiosperm seeds differs between monocots and dicots. In monocots, such as corn and wheat, the single cotyledon is called a scutellum; the scutellum is connected directly to the embryo via vascular tissue (xylem and phloem). Food reserves are stored in the large endosperm. Upon germination, enzymes are secreted by the aleurone, a single layer of cells just inside the seed coat that surrounds the endosperm and embryo. The enzymes degrade the stored carbohydrates, proteins and lipids, the products of which are absorbed by the scutellum and transported via a vasculature strand to the developing embryo. Therefore, the scutellum can be seen to be an absorptive organ, not a storage organ.

The two cotyledons in the dicot seed also have vascular connections to the embryo. In endospermic dicots, the food reserves are stored in the endosperm. During germination, the two cotyledons therefore act as absorptive organs to take up the enzymatically released food reserves, much like in monocots (monocots, by definition, also have endospermic seeds). Tobacco (Nicotiana tabaccum), tomato (Solanum lycopersicum), and pepper (Capsicum annuum) are examples of endospermic dicots. In non-endospermic dicots, the triploid endosperm develops normally following double fertilization, but the endosperm food reserves are quickly remobilized and moved into the developing cotyledon for storage. The two halves of a peanut seed (Arachis hypogaea) and the split peas (Pisum sativum) of split pea soup are individual cotyledons loaded with food reserves.

The seed, along with the ovule, is protected by a seed coat that is formed from the integuments of the ovule sac. In dicots, the seed coat is further divided into an outer coat known as the testa and inner coat known as the tegmen.

The embryonic axis consists of three parts: the plumule, the radicle, and the hypocotyl. The portion of the embryo between the cotyledon attachment point and the radicle is known as the hypocotyl (hypocotyl means "below the cotyledons"). The embryonic axis terminates in a radicle (the embryonic root), which is the region from which the root will develop. In dicots, the hypocotyls extend above ground, giving rise to the stem of the plant. In monocots, the hypocotyl does not show above ground because monocots do not exhibit stem elongation. The part of the embryonic axis that projects above the cotyledons is known as the epicotyl. The plumule is composed of the epicotyl, young leaves, and the shoot apical meristem.

Upon germination in dicot seeds, the epicotyl is shaped like a hook with the plumule pointing downwards. This shape is called the plumule hook, and it persists as long as germination proceeds in the dark. Therefore, as the epicotyl pushes through the tough and abrasive soil, the plumule is protected from damage. Upon exposure to light, the hypocotyl hook straightens out, the young foliage leaves face the sun and expand, and the epicotyl continues to elongate. During this time, the radicle is also growing and producing the primary root. As it grows downward to form the tap root, lateral roots branch off to all sides, producing the typical dicot tap root system.

In monocot seeds (figure), the testa and tegmen of the seed coat are fused. As the seed germinates, the primary root emerges, protected by the root-tip covering: the coleorhiza. Next, the primary shoot emerges, protected by the coleoptile: the covering of the shoot tip. Upon exposure to light (i.e. when the plumule has exited the soil and the protective coleoptile is no longer needed), elongation of the coleoptile ceases and the leaves expand and unfold. At the other end of the embryonic axis, the primary root soon dies, while other, adventitious roots (roots that do not arise from the usual place – i.e. the root) emerge from the base of the stem. This gives the monocot a fibrous root system.

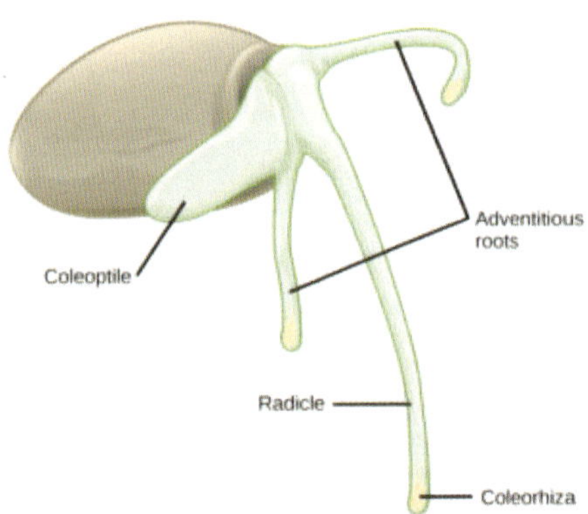

Figure: As this monocot grass seed germinates, the primary root, or radicle, emerges first, followed by the primary shoot, or coleoptile, and the adventitious roots.

Seed Germination

Many mature seeds enter a period of inactivity, or extremely low metabolic activity: a process known as dormancy, which may last for months, years or even centuries. Dormancy helps keep seeds viable during unfavorable conditions. Upon a return to favorable conditions, seed germination takes place. Favorable conditions could be as diverse as moisture, light, cold, fire, or chemical treatments. After heavy rains, many new seedlings emerge. Forest fires also lead to the emergence of new seedlings. Some seeds require vernalization (cold treatment) before they can germinate.

This guarantees that seeds produced by plants in temperate climates will not germinate until the spring. Plants growing in hot climates may have seeds that need a heat treatment in order to germinate, to avoid germination in the hot, dry summers. In many seeds, the presence of a thick seed coat retards the ability to germinate. Scarification, which includes mechanical or chemical processes to soften the seed coat, is often employed before germination. Presoaking in hot water, or passing through an acid environment, such as an animal's digestive tract, may also be employed.

Depending on seed size, the time taken for a seedling to emerge may vary. Species with large seeds have enough food reserves to germinate deep below ground, and still extend their epicotyl all the way to the soil surface. Seeds of small-seeded species usually require light as a germination cue. This ensures the seeds only germinate at or near the soil surface (where the light is greatest). If they were to germinate too far underneath the surface, the developing seedling would not have enough food reserves to reach the sunlight.

Development of Fruit and Fruit Types

After fertilization, the ovary of the flower usually develops into the fruit. Fruits are usually associated with having a sweet taste; however, not all fruits are sweet. Botanically, the term "fruit" is used for a ripened ovary. In most cases, flowers in which fertilization has taken place will develop into fruits, and flowers in which fertilization has not taken place will not. Some fruits develop from the ovary and are known as true fruits, whereas others develop from other parts of the female gametophyte and are known as accessory fruits. The fruit encloses the seeds and the developing embryo, thereby providing it with protection. Fruits are of many types, depending on their origin and texture. The sweet tissue of the blackberry, the red flesh of the tomato, the shell of the peanut, and the hull of corn (the tough, thin part that gets stuck in your teeth when you eat popcorn) are all fruits. As the fruit matures, the seeds also mature.

Fruits may be classified as simple, aggregate, multiple, or accessory, depending on their origin (figure). If the fruit develops from a single carpel or fused carpels of a single ovary, it is known as a simple fruit, as seen in nuts and beans. An aggregate fruit is one that develops from more than one carpel, but all are in the same flower: the mature carpels fuse together to form the entire fruit, as seen in the raspberry. Multiple fruit develops from an inflorescence or a cluster of flowers. An example is the pineapple, where the flowers fuse together to form the fruit. Accessory fruits (sometimes called false fruits) are not derived from the ovary, but from another part of the flower, such as the receptacle (strawberry) or the hypanthium (apples and pears).

Simple fruit | Aggregate fruits | Multiple fruits | Accessory fruit

There are four main types of fruits. Simple fruits, such as these nuts, are derived from a single ovary. Aggregate fruits, like raspberries, form from many carpels that fuse together. Multiple fruits, such as pineapple, form from a cluster of flowers called an inflorescence. Accessory fruit, like the apple, are formed from a part of the plant other than the ovary.

Fruits generally have three parts: the exocarp (the outermost skin or covering), the mesocarp (middle part of the fruit), and the endocarp (the inner part of the fruit). Together, all three are known as the pericarp. The mesocarp is usually the fleshy, edible part of the fruit; however, in some fruits, such as the almond, the endocarp is the edible part. In many fruits, two or all three of the layers are fused, and are indistinguishable at maturity. Fruits can be dry or fleshy. Furthermore, fruits can be divided into dehiscent or indehiscent types. Dehiscent fruits, such as peas, readily release their seeds, while indehiscent fruits, like peaches, rely on decay to release their seeds.

Fruit and Seed Dispersal

The fruit has a single purpose: seed dispersal. Seeds contained within fruits need to be dispersed far from the mother plant, so they may find favorable and less competitive conditions in which to germinate and grow.

Some fruit have built-in mechanisms so they can disperse by themselves, whereas others require the help of agents like wind, water, and animals (figure). Modifications in seed structure, composition, and size help in dispersal. Wind-dispersed fruit are lightweight and may have wing-like appendages that allow them to be carried by the wind. Some have a parachute-like structure to keep them afloat. Some fruits—for example, the dandelion—have hairy, weightless structures that are suited to dispersal by wind.

(a) (b) (c)

Figure: Fruits and seeds are dispersed by various means. (a) Dandelion seeds are dispersed by wind, the (b) coconut seed is dispersed by water, and the (c) acorn is dispersed by animals that cache and then forget it.

Seeds dispersed by water are contained in light and buoyant fruit, giving them the ability to float. Coconuts are well known for their ability to float on water to reach land where they can germinate. Similarly, willow and silver birches produce lightweight fruit that can float on water.

Animals and birds eat fruits, and the seeds that are not digested are excreted in their droppings some distance away. Some animals, like squirrels, bury seed-containing fruits for later use; if the squirrel does not find its stash of fruit, and if conditions are favorable, the seeds germinate. Some fruits, like the cocklebur, have hooks or sticky structures that stick to an animal's coat and are then transported to another place. Humans also play a big role in dispersing seeds when they carry fruits to new places and throw away the inedible part that contains the seeds.

Axeual Reproduction

Vegetative Propagation is an asexual method of reproduction in plants. In plants, the units of vegetative propagation such as runner, rhizome, sucker, tuber, oset, bulb are all capable of giving rise to new ospring. These structures are called vegetative propagules. These structures are called vegetative propagules. Since the formation of these structures does not involve two parents, the process involved is asexual.

Tuberous Roots: The roots of such plants have adventitious buds on their surface which sprout under favourable conditions to produce leafy shoots. In Commercial Production these sprouts are separated and planted. Thus many plants can be obtained from a single root .e.g. Sweet potato, Asparagus (Shatavari), Dahlia.

Stem Tubers: A stem tuber has many notches called 'eyes' on their surface. Each eye actually a node and consists of one or more small axillary buds and reduced scale leaves. After dormancy period the eyes which sprout under favourable conditions to produce leafy shoots. In Commercial Production the tuber is cut into pieces, such that each piece has at least one eye. Then they are grown separately. Thus many plants can be obtained from a single tuber .e.g. Potato, Ginger.

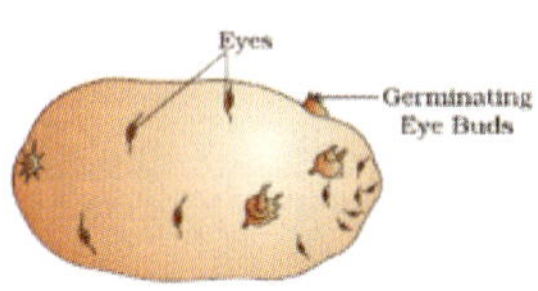

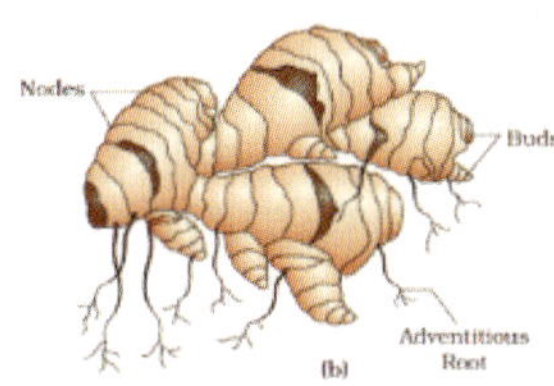

Runners: Runner develops from the lower axillary bud of stem and thin elongated cylindrical wire-like structure with long internodes. It creeps on the ground and becomes rooted at the nodes. Shoots are produced from upper sides of nodes. After getting detached from a parent, such shoots grow as an independent plant .e.g. Cynodon (doobgrass), Fragaria (strawberry), Oxalis.

Leaf: In bryophyllum, the leaf is succulent with crenate or notched margins. Adventitious buds called epiphyllous buds or foliar buds are formed at notches at the tip of lateral veins. These buds sprout and form leafy shoots and adventitious roots. When such sprout falls on wet soil, they develop into independent plants. e.g. Bryophyllum, Kalanchoe, Begonia.

Hazard due to Water Hyacinth (Terror of Bengal): One of the most invasive weeds. Grows wherever there is standing water. Drains oxygen from water-leads to the death of fishes. Introduced in India because of its pretty flowers & shape of leaves. Vegetative propagation occurs at a phenomenal rate.

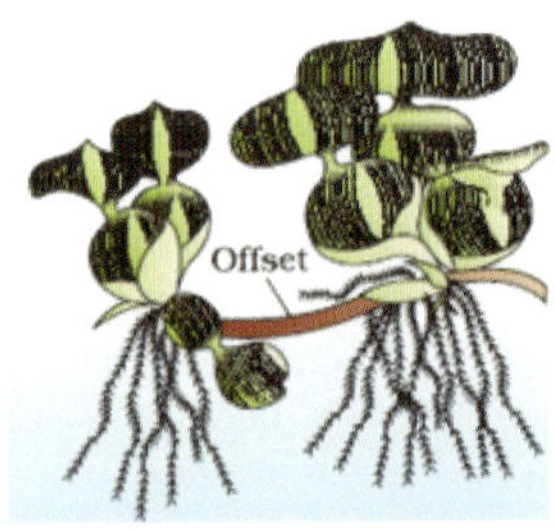

Artificial Methods of Vegetative Propagation (Asexual reproduction):

Cutting: In this method, a branch of plant part (stem, root or leaves) is cut with a node (primary meristem) on it. This piece of a branch is termed a cutting. This cutting is buried in the soil. The cutting is watered continuously. This method is the cheapest and convenient method of vegetative propagation.

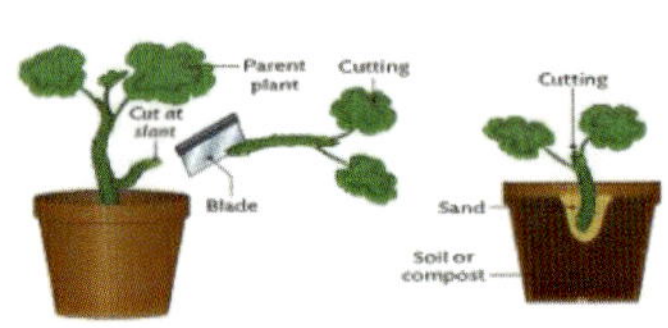

Graing: In this method, the parts of two dierent plants are joined together in such a way that they unite and continue their growth as one plant. The plant rooted to the soil and on which the part of the other plant is inserted is called a stock .The other plant which is inserted on the stock is called Scion or graft. Budding is a special case of graing in which a single bid with a small part of a bark and living tissue is grafted on another plant.

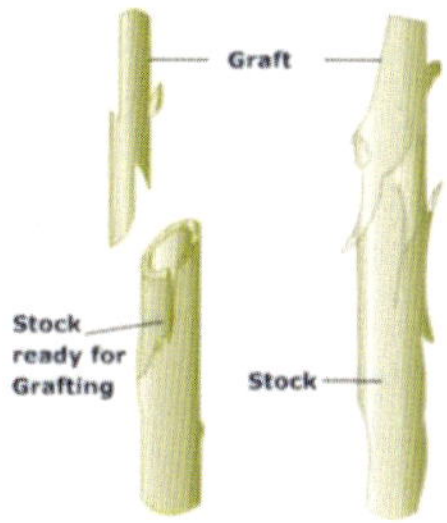

The Significance of Vegetative Propagation:

It is easy and cheaper method of multiplication . The plants like banana, pineapple and grapes which do not produce viable seeds can be propagated by vegetative propagation only. In this method, genetically similar plants are formed. The yield can be increased by graing high-yield variety on less yielding variety but which is adapted to the region. This is a rapid method of propagation particularly for the plants having a long dormancy period. It can be used to get clones of rare plants.

Animal Reproduction

Some animals produce offspring through asexual reproduction while other animals produce offspring through sexual reproduction. Both methods have advantages and disadvantages. Asexual reproduction produces offspring that are genetically identical to the parent because the offspring are all clones of the original parent. A single individual can produce offspring asexually and large numbers of offspring can be produced quickly; these are two advantages that asexually reproducing organisms have over sexually reproducing organisms. In a stable or predictable environment, asexual reproduction is an effective means of reproduction because all the offspring will be adapted to that environment. In an unstable or unpredictable environment, species that reproduce asexually may be at a disadvantage because all the offspring are genetically identical and may not be adapted to different conditions.

During sexual reproduction, the genetic material of two individuals is combined to produce genetically diverse offspring that differ from their parents. The genetic diversity of sexually produced offspring is thought to give sexually reproducing individuals greater fitness because more of their offspring may survive and reproduce in an unpredictable or changing environment. Species that reproduce sexually (and have separate sexes) must maintain two different types of individuals, males and females. Only half the population (females) can produce the offspring, so fewer offspring will be produced when compared to asexual reproduction. This is a disadvantage of sexual reproduction compared to asexual reproduction.

Asexual Reproduction

Asexual reproduction occurs in prokaryotic microorganisms (bacteria and archaea) and in many eukaryotic, single-celled and multi-celled organisms. There are several ways that animals reproduce asexually, the details of which vary among individual species.

Fission

Fission, also called binary fission, occurs in some invertebrate, multi-celled organisms. It is in some ways analogous to the process of binary fission of single-celled prokaryotic organisms. The term fission is applied to instances in which an organism appears to split itself into two parts and, if necessary, regenerate the missing parts of each new organism. For example, species of turbellarian flatworms commonly called the planarians, such as Dugesia dorotocephala, are able to separate their bodies into head and tail regions and then regenerate the missing half in each of the two new organisms. Sea anemones (Cnidaria), such as species of the genus Anthopleura, will divide along the oral-aboral axis, and sea cucumbers (Echinodermata) of the genus Holothuria, will divide into two halves across the oral-aboral axis and regenerate the other half in each of the resulting individuals.

Figure: The Anthopleura artemisia sea anemone can reproduce through fission.

Budding

Budding is a form of asexual reproduction that results from the outgrowth of a part of the body leading to a separation of the "bud" from the original organism and the formation of two individuals, one smaller than the other. Budding occurs commonly in some invertebrate animals such as hydras and corals. In hydras, a bud forms that develops into an adult and breaks away from the main body (figure).

Figure (a) Hydra reproduce asexually through budding: a bud forms on the tubular body of an adult hydra, develops a mouth and tentacles, and then detaches from its parent. The new hydra is fully developed and will find its own location for attachment. (b) Some coral, such as the Lophelia pertusa shown here, can reproduce through budding. Part a: This shows a hydra, which has a stalk-like body with tentacles growing out the top. A smaller hydra is budding from the side of the stalk. Part b: This photo shows branching white coral polyps.

Fragmentation

Fragmentation is the breaking of an individual into parts followed by regeneration. If the animal is capable of fragmentation, and the parts are big enough, a separate individual will regrow from each part. Fragmentation may occur through accidental damage, damage from predators, or as a natural form of reproduction. Reproduction through fragmentation is observed in sponges, some cnidarians, turbellarians, echinoderms, and annelids. In some sea stars, a new individual can be regenerated from a broken arm and a piece of the central disc. This sea star (figure) is in the process of growing a complete sea star from an arm that has been cut off. Fisheries workers have been known to try to kill the sea stars eating their clam or oyster beds by cutting them in half and throwing them back into the ocean. Unfortunately for the workers, the two parts can each regenerate a new half, resulting in twice as many sea stars to prey upon the oysters and clams.

Figure: (a) Linckia multifora is a species of sea star that can reproduce asexually via fragmentation. In this process, (b) an arm that has been shed grows into a new sea star.

Parthenogenesis

Parthenogenesis is a form of asexual reproduction in which an egg develops into an individual without being fertilized. The resulting offspring can be either haploid or diploid, depending on the process in the species. Parthenogenesis occurs in invertebrates such as water fleas, rotifers, aphids, stick insects, and ants, wasps, and bees. Ants, bees, and wasps use parthenogenesis to produce haploid males (drones). The diploid females (workers and queens) are the result of a fertilized egg.

Some vertebrate animals—such as certain reptiles, amphibians, and fish—also reproduce through parthenogenesis. Parthenogenesis has been observed in species in which the sexes were separated in terrestrial or marine zoos. Two female Komodo dragons, a hammerhead shark, and a blacktop shark have produced parthenogenic young when the females have been isolated from males. It is possible that the asexual reproduction observed occurred in response to unusual circumstances and would normally not occur.

Sexual Reproduction

Sexual reproduction is the combination of reproductive cells from two individuals to form genetically unique offspring. The nature of the individuals that produce the two kinds of gametes can vary, having for example separate sexes or both sexes in each individual. Sex determination, the mechanism that determines which sex an individual develops into, also can vary.

Hermaphroditism

Hermaphroditism occurs in animals in which one individual has both male and female reproductive systems. Invertebrates such as earthworms, slugs, tapeworms, and snails (figure) are often hermaphroditic. Hermaphrodites may self-fertilize, but typically they will mate with another of their species, fertilizing each other and both producing offspring. Self-fertilization is more common in animals that have limited mobility or are not motile, such as barnacles and clams. Many species have specific mechanisms in place to prevent self-fertilization, because it is an extreme form of inbreeding and usually produces less fit offspring.

(a) (b)

Figure: Many (a) snails are hermaphrodites. When two individuals (b) mate, they can produce up to 100 eggs each.

Sex Determination

Mammalian sex is determined genetically by the combination of X and Y chromosomes. Individuals homozygous for X (XX) are female and heterozygous individuals (XY) are male. In mammals, the presence of a Y chromosome causes the development of male characteristics and its absence results in female characteristics. The XY system is also found in some insects and plants.

Bird sex determination is dependent on the combination of Z and W chromosomes. Homozygous for Z (ZZ) results in a male and heterozygous (ZW) results in a female. Notice that this system is the opposite of the mammalian system because in birds the female is the sex with the different sex chromosomes. The W appears to be essential in determining the sex of the individual, similar to the Y chromosome in mammals. Some fish, crustaceans, insects (such as butterflies and moths), and reptiles use the ZW system.

More complicated chromosomal sex determining systems also exist. For example, some swordtail fish have three sex chromosomes in a population.

The sex of some other species is not determined by chromosomes, but by some aspect of the environment. Sex determination in alligators, some turtles, and tuataras, for example, is dependent on the temperature during the middle third of egg development. This is referred to as environmental sex determination, or more specifically, as temperature-dependent sex determination. In many turtles, cooler temperatures during egg incubation produce males and warm temperatures produce females, while in many other species of turtles, the reverse is true. In some crocodiles and some turtles, moderate temperatures produce males and both warm and cool temperatures produce females.

Individuals of some species change their sex during their lives, switching from one to the other. If the individual is female first, it is termed protogyny or "first female," if it is male first, it is termed protandry or "first male." Oysters are born male, grow in size, and become female and lay eggs. The wrasses, a family of reef fishes, are all sequential hermaphrodites. Some of these species live in closely coordinated schools with a dominant male and a large number of smaller females. If the male dies, a female increases in size, changes sex, and becomes the new dominant male.

Fertilization

The fusion of a sperm and an egg is a process called fertilization. This can occur either inside (internal fertilization) or outside (external fertilization) the body of the female. Humans provide an example of the former, whereas frog reproduction is an example of the latter.

External Fertilization

External fertilization usually occurs in aquatic environments where both eggs and sperm are released into the water. After the sperm reaches the egg, fertilization takes place. Most external fertilization happens during the process of spawning where one or several females release their eggs and the male(s) release sperm in the same area, at the same time. The spawning may be triggered by environmental signals, such as water temperature or the length of daylight. Nearly all fish spawn, as do crustaceans (such as crabs and shrimp), mollusks (such as oysters), squid, and echinoderms (such as sea urchins and sea cucumbers). Frogs, corals, mayflies, and mosquitoes also spawn (figure).

Figure: During sexual reproduction in toads, the male grasps the female from behind and externally fertilizes the eggs as they are deposited.

Internal Fertilization

Internal fertilization occurs most often in terrestrial animals, although some aquatic animals also use this method. Internal fertilization may occur by the male directly depositing sperm in the female during mating. It may also occur by the male depositing sperm in the environment, usually in a protective structure, which a female picks up to deposit the sperm in her reproductive tract. There are three ways that offspring are produced following internal fertilization. In oviparity, fertilized eggs are laid outside the female's body and develop there, receiving nourishment from the yolk that is a part of the egg (figure(a)). This occurs in some bony fish, some reptiles, a few cartilaginous fish, some amphibians, a few mammals, and all birds. Most non-avian reptiles and insects produce leathery eggs, while birds and some turtles produce eggs with high concentrations of calcium carbonate in the shell, making them hard. Chicken eggs are an example of a hard shell. The eggs of the egg-laying mammals such as the platypus and echidna are leathery.

In ovoviparity, fertilized eggs are retained in the female, and the embryo obtains its nourishment from the egg's yolk. The eggs are retained in the female's body until they hatch inside of her, or she lays the eggs right before they hatch. This process helps protect the eggs until hatching. This occurs in some bony fish (like the platyfish Xiphophorus maculatus, figure (b)), some sharks, lizards, some snakes (garter snake Thamnophis sirtalis), some vipers, and some invertebrate animals (Madagascar hissing cockroach Gromphadorhina portentosa).

In viviparity the young are born alive. They obtain their nourishment from the female and are born in varying states of maturity. This occurs in most mammals (figure (c)), some cartilaginous fish, and a few reptiles.

(a)

(b)

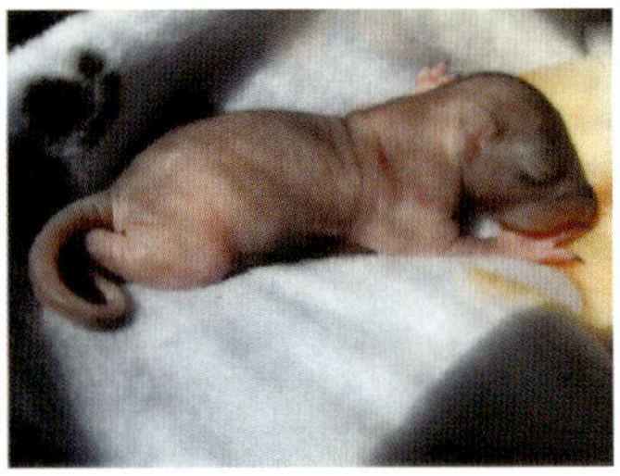

(c)

In (a) oviparity, young develop in eggs outside the female body, as with these Harmonia axydridis beetles hatching. Some aquatic animals, like this (b) pregnant Xiphophorus maculatus are ovoviparous, with the egg developing inside the female and nutrition supplied primarily from the yolk. In mammals, nutrition is supported by the placenta, as was the case with this (c) newborn squirrel.

Human Reproduction System

Human reproductive system is an organ system by which humans reproduce and bear live offspring. Provided all organs are present, normally constructed, and functioning properly, the essential features of human reproduction are (1) liberation of an ovum, or egg, at a specific time in the reproductive cycle, (2) internal fertilization of the ovum by spermatozoa, or sperm cells, (3) transport of the fertilized ovum to the uterus, or womb, (4) implantation of the blastocyst, the early embryo developed from the fertilized ovum, in the wall of the uterus, (5) formation of a placenta and maintenance of the unborn child during the entire period of gestation, (6) birth of the child and expulsion of the placenta, and (7) suckling and care of the child, with an eventual return of the maternal organs to virtually their original state.

For this biological process to be carried out, certain organs and structures are required in both the male and the female. The source of the ova (the female germ cells) is the female ovary; that of spermatozoa (the male germ cells) is the testis. In females, the two ovaries are situated in the pelvic cavity; in males, the two testes are enveloped in a sac of skin, the scrotum, lying below and outside the abdomen. Besides producing the germ cells, or gametes, the ovaries and testes are the source of hormones that cause full development of secondary sexual characteristics and also the proper functioning of the reproductive tracts. These tracts comprise the fallopian tubes, the uterus, the vagina, and associated structures in females and the penis, the sperm channels (epididymis, ductus deferens, and ejaculatory ducts), and other related structures and glands in males. The function of the fallopian tube is to convey an ovum, which is fertilized in the tube, to the uterus, where gestation (development before birth) takes place. The function of the male ducts is to convey spermatozoa from the testis, to store them, and, when ejaculation occurs, to eject them with secretions from the male glands through the penis.

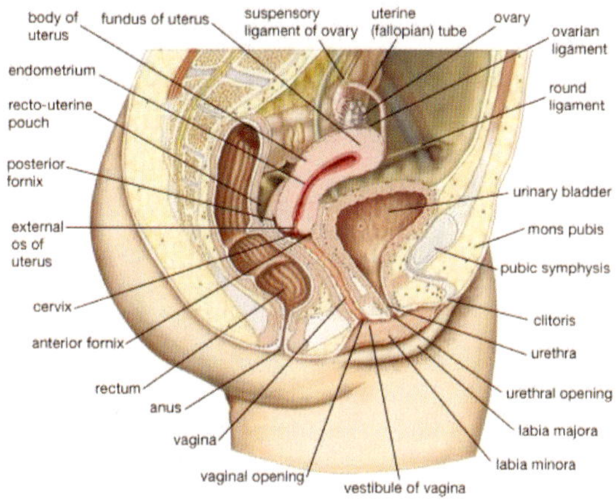

At copulation, or sexual intercourse, the erect penis is inserted into the vagina, and spermatozoa contained in the seminal fluid (semen) are ejaculated into the female genital tract. Spermatozoa then pass from the vagina through the uterus to the fallopian tube to fertilize the ovum in the outer part of the tube. Females exhibit a periodicity in the activity of their ovaries and uterus, which starts at puberty and ends at the menopause. The periodicity is manifested by menstruation at intervals of about 28 days; important changes occur in the ovaries and uterus during each reproductive, or menstrual, cycle. Periodicity, and subsequently menstruation, is suppressed during pregnancy and lactation.

Development of The Reproductive Organs

The sex of a child is determined at the time of fertilization of the ovum by the spermatozoon. The differences between a male and a female are genetically determined by the chromosomes that each possesses in the nuclei of the cells. Once the genetic sex has been determined, there normally follows a succession of changes that will result, finally, in the development of an adult male or female. There is, however, no external indication of the sex of an embryo during the first eight weeks of its life within the uterus. This is a neutral or indifferent stage during which the sex of an embryo can be ascertained only by examination of the chromosomes in its cells.

The next phase, one of differentiation, begins first in gonads that are to become testes and a week or so later in those destined to be ovaries. Embryos of the two sexes are initially alike in possessing similar duct systems linking the undifferentiated gonads with the exterior and in having similar external genitalia, represented by three simple protuberances. The embryos each have four ducts, the subsequent fate of which is of great significance in the eventual anatomical differences between men and women. Two ducts closely related to the developing urinary system are called mesonephric, or wolffian, ducts. In males each mesonephric duct becomes differentiated into four related structures: a duct of the epididymis, a ductus deferens, an ejaculatory duct, and a seminal vesicle. In females the mesonephric ducts are largely suppressed. The other two ducts, called the paramesonephric or müllerian ducts, persist, in females, to develop into the fallopian tubes, the uterus, and part of the vagina; in males they are largely suppressed. Differentiation also occurs in the primitive external genitalia, which in males become the penis and scrotum and in females the vulva (the clitoris, labia, and vestibule of the vagina).

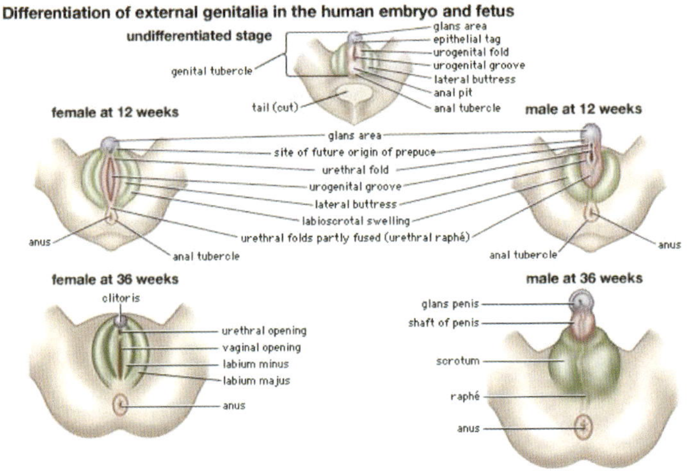

At birth the organs appropriate to each sex have developed and are in their adult positions but are not functioning. Various abnormalities can occur during development of sex organs in embryos, leading to hermaphroditism, pseudohermaphroditism, and other chromosomally induced conditions. During childhood until puberty there is steady growth in all reproductive organs and a gradual development of activity. Puberty marks the onset of increased activity in the sex glands and the steady development of secondary sexual characteristics.

In males at puberty the testes enlarge and become active, the external genitalia enlarge, and the capacity to ejaculate develops. Marked changes in height and weight occur as hormonal secretion from the testes increases. The larynx, or voice box, enlarges, with resultant deepening of the voice. Certain features in the skeleton, as seen in the pelvic bones and skull, become accentuated. The hair in the armpits and the pubic hair becomes abundant and thicker. Facial hair develops, as well as hair on the chest, abdomen, and limbs. Hair at the temples recedes. Skin glands become more active, especially apocrine glands (a type of sweat gland that is found in the armpits and groin and around the anus).

In females at puberty, the external genitalia enlarge and the uterus commences its periodic activity with menstruation. The breasts develop, and there is a deposition of body fat in accordance with the usual contours of the mature female. Growth of axillary (armpit) and pubic hair is more abundant, and the hair becomes thicker.

The Male Reproductive System

The male gonads are the testes; they are the source of spermatozoa and also of male sex hormones called androgens. The other genital organs are the epididymides; the ductus, or vasa, deferentia; the seminal vesicles; the ejaculatory ducts; and the penis; as well as certain accessory structures, such as the prostate and the bulbourethral (Cowper) glands. The principal functions of these structures are to transport the spermatozoa from the testes to the exterior, to allow their maturation on the way, and to provide certain secretions that help form the semen.

External Genitalia

The Penis

The penis, the male organ of copulation, is partly inside and partly outside the body. The inner part, attached to the bony margins of the pubic arch (that part of the pelvis directly in front and at the base of the trunk), is called the root of the penis. The second, or outer, portion is free, pendulous, and enveloped all over in skin; it is termed the body of the penis. The organ is composed chiefly of cavernous or erectile tissue that becomes engorged with blood to produce considerable enlargement and erection. The penis is traversed by a tube, the urethra, which serves as a passage both for urine and for semen.

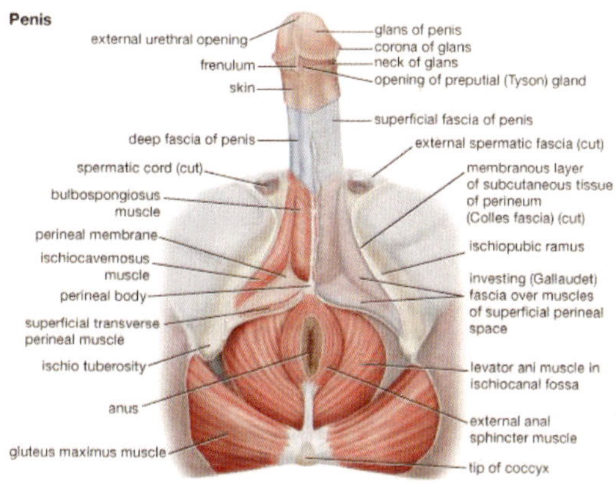

The human penis.

The body of the penis, sometimes referred to as the shaft, is cylindrical in shape when flaccid but when erect is somewhat triangular in cross section, with the angles rounded. This condition arises because the right corpus cavernosum and the left corpus cavernosum, the masses of erectile tissue, lie close together in the dorsal part of the penis, while a single body, the corpus spongiosum, which contains the urethra, lies in a midline groove on the undersurface of the corpora cavernosa. The dorsal surface of the penis is that which faces upward and backward during erection.

The slender corpus spongiosum reaches beyond the extremities of the erectile corpora cavernosa and at its outer end is enlarged considerably to form a soft, conical, sensitive structure called the glans penis. The base of the glans has a projecting margin, the corona, and the groove where the corona overhangs the corpora cavernosa is referred to as the neck of the penis. The glans is traversed by the urethra, which ends in a vertical, slitlike, external opening. The skin over the penis is thin and loosely adherent and at the neck is folded forward over the glans for a variable distance to form the prepuce or foreskin. A median fold, the frenulum of the prepuce, passes to the undersurface of the glans to reach a point just behind the urethral opening. The prepuce can usually be readily drawn back to expose the glans.

The root of the penis comprises two crura, or projections, and the bulb of the penis. The crura and the bulb are attached respectively to the edges of the pubic arch and to the perineal membrane (the fibrous membrane that forms a floor of the trunk). Each crus is an elongated structure covered by the ischiocavernosus muscle, and each extends forward, converging toward the other, to become continuous with one of the corpora cavernosa. The oval bulb of the penis lies between the two crura and is covered by the bulbospongiosus muscle. It is continuous with the corpus spongiosum. The urethra enters it on the flattened deep aspect that lies against the perineal membrane, traverses its substances, and continues into the corpus spongiosum.

The two corpora cavernosa are close to one another, separated only by a partition in the fibrous sheath that encloses them. The erectile tissue of the corpora is divided by numerous small fibrous bands into many cavernous spaces, relatively empty when the penis is flaccid but engorged with blood during erection. The structure of the tissue of the corpus spongiosum is similar to that of the corpora cavernosa, but there is more smooth muscle and elastic tissue. A deep fascia, or sheet of connective tissue, surrounding the structures in the body of the penis is prolonged to form the suspensory ligament, which anchors the penis to the pelvic bones at the midpoint of the pubic arch.

The penis has a rich blood supply from the internal pudendal artery, a branch of the internal iliac artery, which supplies blood to the pelvic structures and organs, the buttocks, and the inside of the thighs. Erection is brought about by distension of the cavernous spaces with blood, which is prevented from draining away by compression of the veins in the area.

The penis is amply supplied with sensory and autonomic (involuntary) nerves. Of the autonomic nerve fibres the sympathetic fibres cause constriction of blood vessels, and the parasympathetic fibres cause their dilation. It is usually stated that ejaculation is brought about by the sympathetic system, which at the same time inhibits the desire to urinate and also prevents the semen from entering the bladder.

The Scrotum

The scrotum is a pouch of skin lying below the pubic symphysis and just in front of the upper parts of the thighs. It contains the testes and lowest parts of the spermatic cord. A scrotal septum or partition divides the pouch into two compartments and arises from a ridge, or raphe, visible on the outside of the scrotum. The raphe turns forward onto the undersurface of the penis and is continued back onto the perineum (the area between the legs and as far back as the anus). This arrangement indicates the bilateral origin of the scrotum from two genital swellings that lie one on each side of the base of the phallus, the precursor of the penis or clitoris in the embryo. The swellings are also referred to as the labioscrotal swellings, because in females they remain separate to form the labia majora and in males they unite to form the scrotum.

The skin of the scrotum is thin, pigmented, devoid of fatty tissue, and more or less folded and wrinkled. There are some scattered hairs and sebaceous glands on its surface. Below the skin is a layer of involuntary muscle, the dartos, which can alter the appearance of the scrotum. On exposure of the scrotum to cold air or cold water, the dartos contracts and gives the scrotum a shortened, corrugated appearance; warmth causes the scrotum to become smoother, flaccid, and less closely tucked in around the testes. Beneath the dartos muscle are layers of fascia continuous with those forming the coverings of each of the two spermatic cords, which suspend the testes within the scrotum and contain each ductus deferens, the testicular blood and lymph vessels, the artery to the cremaster muscle (which draws the testes upward), the artery to each ductus deferens, the genital branch of the genitofemoral nerve, and the testicular network of nerves.

The scrotum is supplied with blood by the external pudendal branches of the femoral artery, which is the chief artery of the thigh, and by the scrotal branches of the internal pudendal artery. The veins follow the arteries. The lymphatic drainage is to the lymph nodes in the groin.

The Testes

The two testes, or testicles, which usually complete their descent into the scrotum from their point of origin on the back wall of the abdomen in the seventh month after conception, are suspended in the scrotum by the spermatic cords. Each testis is 4 to 5 cm (about 1.5 to 2 inches) long and is enclosed in a fibrous sac, the tunica albuginea. This sac is lined internally by the tunica vasculosa, containing a network of blood vessels, and is covered by the tunica vaginalis, which is a continuation of the membrane that lines the abdomen and pelvis. The tunica albuginea has extensions into each testis that act as partial partitions to divide the testis into approximately 250 compartments, or lobules.

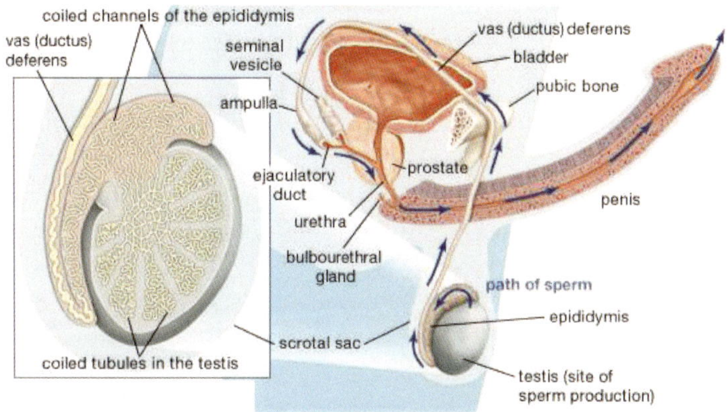

Structures involved in the production and transport of semen.

Each lobule contains one or more convoluted tubules, or narrow tubes, where sperm are formed. The tubules, if straightened, would extend about 70 cm (about 28 inches). The multistage process of sperm formation, which takes about 60 days, goes on in the lining of the tubules, starting with the spermatogonia, or primitive sperm cells, in the outermost layer of the lining. Spermatozoa (sperm) leaving the tubules are not capable of independent motion, but they undergo a further maturation process in the ducts of the male reproductive tract; the process may be continued when, after ejaculation, they pass through the female tract. Maturation of the sperm in the female tract is called capacitation.

Each spermatozoon is a slender elongated structure with a head, a neck, a middle piece, and a tail. The head contains the cell nucleus. When the spermatozoon is fully mature, it is propelled by the lashing movements of the tail.

The male sex hormone testosterone is produced by Leydig cells. These cells are located in the connective (interstitial) tissue that holds the tubules together within each lobule. The tissue becomes markedly active at puberty under the influence of the interstitial-cell-stimulating hormone of the anterior lobe of the pituitary gland; this hormone in women is called luteinizing hormone. Testosterone stimulates the male accessory sex glands (prostate, seminal vesicles) and also brings about the development of male secondary sex characteristics at puberty. The hormone may also be necessary to cause maturation of sperm and to heighten the sex drive of the male. The testis is also the source of some of the female sex hormone estrogen, which may exert an influence on pituitary activity.

Each testis is supplied with blood by the testicular arteries, which arise from the front of the aorta just below the origin of the renal (kidney) arteries. Each artery crosses the rear abdominal wall, enters the spermatic cord, passes through the inguinal canal, and enters the upper end of each testis at the back. The veins leaving the testis and epididymis form a network, which ascends into the spermatic cord. The lymph vessels, which also pass through the spermatic cord, drain to the lateral and preaortic lymph nodes. Nerve fibres to the testis accompany the vessels; they pass through the renal and aortic nerve plexuses, or networks.

Structures of the Sperm Canal

The epididymis, ductus deferens (or vas deferens), and ejaculatory ducts form the sperm canal. Together they extend from the testis to the urethra, where it lies within the prostate. Sperm are conveyed from the testis along some 20 ductules, or small ducts, which pierce the fibrous capsule to enter the head of the epididymis. The ductules are straight at first but become dilated and then much convoluted to form distinct compartments within the head of the epididymis. They each open into a single duct, the highly convoluted duct of the epididymis, which constitutes the "body" and "tail" of the structure. It is held together by connective tissue but if unraveled would be nearly 6 metres (20 feet) long. The duct enlarges and becomes thicker-walled at the lower end of the tail of the epididymis, where it becomes continuous with the ductus deferens.

The ductules from the testis have a thin muscular coat and a lining that consists of alternating groups of high columnar cells with cilia (hairlike projections) and low cells lacking cilia. The cilia assist in moving sperm toward the epididymis. In the duct of the epididymis the muscle coat is thicker and the lining is thick with tufts of large nonmotile cilia. There is some evidence that the ductules and the first portion of the duct of the epididymis remove excess fluid and extraneous debris from the testicular secretions entering these tubes. The blood supply to the epididymis is by a branch from the testicular artery given off before that vessel reaches the testis.

The ductus deferens, or vas deferens, is the continuation of the duct of the epididymis. It commences at the lower part of the tail of the epididymis and ascends along the back border of the testis to its upper pole. Then, as part of the spermatic cord, it extends to the deep inguinal ring. Separating from the other elements of the spermatic cord—the blood vessels, nerves, and lymph vessels—at the ring, the ductus deferens makes its way through the pelvis toward the base of the prostate, where it is joined by the seminal vesicle to form the ejaculatory duct. A part of the ductus that is dilated and rather tortuous, near the base of the urinary bladder, is called the ampulla.

The ductus deferens has a thick coat of smooth muscle that gives it a characteristic cordlike feel. The longitudinal muscle fibres are well developed, and peristaltic contractions (contractions in waves) move the sperm toward the ampulla. The mucous membrane lining the interior is in longitudinal folds and is mostly covered with nonciliated columnar cells, although some cells have nonmotile cilia. The ampulla is thinner-walled and probably acts as a sperm store.

Accessory Organs

The Prostate Gland, Seminal Vesicles, and Bulbourethral Glands

These structures provide secretions to form the bulk of the seminal fluid of an ejaculate. The prostate gland is in the lesser or true pelvis, centred behind the lower part of the pubic arch. It lies in front of the rectum. The prostate is shaped roughly like an inverted pyramid; its base is directed upward and is immediately continuous with the neck of the urinary bladder. The urethra traverses its substance. The two ejaculatory ducts enter the prostate near the upper border of its posterior surface. The prostate is of a firm consistency, surrounded by a capsule of fibrous tissue and smooth muscle. It measures about 4 cm across, 3 cm in height, and 2 cm front to back (about 1.6 by 1.2 by 0.8 inch) and consists of glandular tissue contained in a muscular framework. It is imperfectly divided into three lobes. Two lobes at the side form the main mass and are continuous behind the urethra. In front of the urethra they are connected by an isthmus of fibromuscular tissue devoid of glands. The third, or median, lobe is smaller and variable in size and may lack glandular tissue. There are three clinically significant concentric zones of prostatic glandular tissue about the urethra. A group of short glands that are closest to the urethra and discharge mucus into its channel are subject to simple enlargement. Outside these is a ring of submucosal glands (glands from which the mucosal glands develop), and farther out is a large outer zone of long branched glands, composing the bulk of the glandular tissue. Prostate cancer is almost exclusively confined to the outer zone. The glands of the outer zone are lined by tall columnar cells that secrete prostatic fluid under the influence of androgens from the testis. The fluid is thin, milky, and slightly acidic.

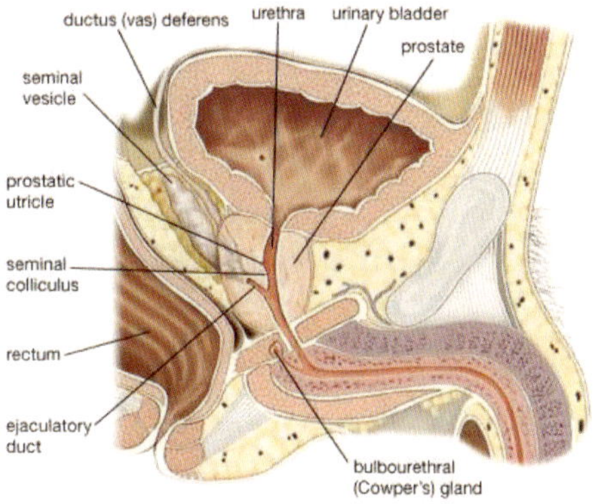

Sagittal section of the male reproductive organs, showing the prostate gland and seminal vesicles.

The seminal vesicles are two structures, about 5 cm (2 inches) in length, lying between the rectum and the base of the bladder. Their secretions form the bulk of semen. Essentially, each vesicle consists of a much-coiled tube with numerous diverticula or outpouches that extend from the main tube, the whole being held together by connective tissue. At its lower end the tube is constricted to form a straight duct or tube that joins with the corresponding ductus deferens to form the ejaculatory duct. The vesicles are close together in their lower parts, but they are separated above where they lie close to the deferent ducts. The seminal vesicles have longitudinal and circular layers of smooth muscle, and their cavities are lined with mucous membrane, which is the source of the secretions of the organs. These secretions are ejected by muscular contractions during ejaculation. The activity of the vesicles is dependent on the production of the hormone androgen by the testes. The secretion is thick, sticky, and yellowish; it contains the sugar fructose and is slightly alkaline.

The bulbourethral glands, often called Cowper glands, are pea-shaped glands that are located beneath the prostate gland at the beginning of the internal portion of the penis. The glands, which measure only about 1 cm (0.4 inch) in diameter, have slender ducts that run forward and toward the centre to open on the floor of the spongy portion of the urethra. They are composed of a network of small tubes, or tubules, and saclike structures; between the tubules are fibres of muscle and elastic tissue that give the glands muscular support. Cells within the tubules and sacs contain droplets of mucus, a thick protein compound. The fluid excreted by these glands is clear and thick and acts as a lubricant; it is also thought to function as a flushing agent that washes out the urethra before the semen is ejaculated; it may also help to make the semen less watery and to provide a suitable living environment for the sperm.

Ejaculatory Ducts

The two ejaculatory ducts lie on each side of the midline and are formed by the union of the duct of the seminal vesicle, which contributes secretions to the semen, with the end of the ductus deferens at the base of the prostate. Each duct is about 2 cm (about 0.8 inch) long and passes between a lateral and the median lobe of the prostate to reach the floor of the prostatic urethra. This part of the urethra has on its floor (or posterior wall) a longitudinal ridge called the urethral crest. On each side is a depression, the prostatic sinus, into which open the prostatic ducts. In the middle of the urethral crest is a small elevation, the colliculus seminalis, on which the opening of the prostatic utricle is found. The prostatic utricle is a short diverticulum or pouch lined by mucous membrane; it may correspond to the vagina or uterus in the female. The small openings of the ejaculatory ducts lie on each side of or just within the opening of the prostatic utricle. The ejaculatory ducts are thin-walled and lined by columnar cells.

The Female Reproductive System

The female gonads, or sexual glands, are the ovaries; they are the source of ova (eggs) and of the female sex hormones estrogens and progestogens. The fallopian, or uterine, tubes conduct ova to the uterus, which lies within the lesser or true pelvis. The uterus connects through the cervical canal with the vagina. The vagina opens into the vestibule about which lie the external genitalia, collectively known as the vulva.

External Genitalia

The female external genitalia include the structures placed about the entrance to the vagina and external to the hymen, the membrane across the entrance to the vagina. They are the mons pubis (also called the mons veneris), the labia majora and minora, the clitoris, the vestibule of the vagina, the bulb of the vestibule, and the greater vestibular glands.

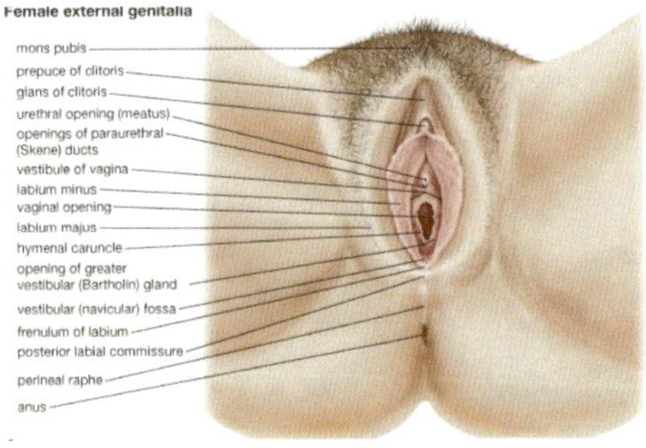

The female external genitalia.

The mons pubis is the rounded eminence, made by fatty tissue beneath the skin, lying in front of the pubic symphysis. A few fine hairs may be present in childhood; later, at puberty, they become coarser and more numerous. The upper limit of the hairy region is horizontal across the lower abdomen.

The labia majora are two marked folds of skin that extend from the mons pubis downward and backward to merge with the skin of the perineum. They form the lateral boundaries of the vulval or pudendal cleft, which receives the openings of the vagina and the urethra. The outer surface of each labium is pigmented and hairy; the inner surface is smooth but possesses sebaceous glands. The labia majora contain fat and loose connective tissue and sweat glands. They correspond to the scrotum in the male and contain tissue resembling the dartos muscle. The round ligament ends in the tissue of the labium. The labia minora are two small folds of skin, lacking fatty tissue, that extend backward on each side of the opening into the vagina. They lie inside the labia majora and are some 4 cm (about 1.5 inches) in length. In front, an upper portion of each labium minus passes over the clitoris—the structure in the female corresponding to the penis (excluding the urethra) in the male—to form a fold, the prepuce of the clitoris, and a lower portion passes beneath the clitoris to form its frenulum. The two labia minora are joined at the back across the midline by a fold that becomes stretched at childbirth. The labia minora lack hairs but possess sebaceous and sweat glands.

The clitoris is a small erectile structure composed of two corpora cavernosa separated by a partition. Partially concealed beneath the forward ends of the labia minora, it possesses a sensitive tip of spongy erectile tissue, the glans clitoridis. The external opening of the urethra is some 2.5 cm (about 1 inch) behind the clitoris and immediately in front of the vaginal opening.

The vestibule of the vagina is the cleft between the labia minora into which the urethra and vagina open. The hymen vaginae lies at the opening of the vagina: it is a thin fold of mucous membrane that varies in shape. After rupture of the hymen, the small rounded elevations that remain are known as the carunculae hymenales. The bulb of the vestibule, corresponding to the bulb of the penis, is two elongated masses of erectile tissue that lie one on each side of the vaginal opening. At their posterior ends lie the greater vestibular glands, small mucous glands that open by a duct in the groove between the hymen and each labium minus. They correspond to the bulbourethral glands of the male.

The blood supply and nerve supply of the female external genital organs are similar to those supplying corresponding structures in the male.

Internal Structures

The vagina

The vagina (the word means "sheath") is the canal that extends from the cervix (outer end) of the uterus within the lesser pelvis down to the vestibule between the labia minora. The orifice of the vagina is guarded by the hymen. The vagina lies behind the bladder and urethra and in front of the rectum and anal canal. Its walls are collapsed; the anterior wall is some 7.5 cm (3 inches) in length, whereas the posterior wall is about 1.5 cm (0.6 inch) longer. The vagina is directed obliquely upward and backward. The axis of the vagina forms an angle of over 90° with that of the uterus. This angle varies considerably depending on conditions in the bladder, in the rectum, and during pregnancy. The cervix of the uterus projects for a short distance into the vagina and is normally pressed against its posterior wall. There are, therefore, recesses in the vagina at the back, on each side, and at the front of the cervix. These are known as the posterior fornix (behind the cervix and the largest), the lateral fornices (at the sides), and the anterior fornix (at the front of the cervix). The position of the uterus in relation to the vagina is described further in the section on the uterus.

The upper part of the posterior wall of the vagina is covered by peritoneum or membrane that is folded back onto the rectum to form the recto-uterine pouch. The lower part of the posterior vaginal wall is separated from the anal canal by a mass of tissue known as the perineal body.

The vagina has a mucous membrane and an outer smooth muscle coat closely attached to it. The mucous membrane has a longitudinal ridge in the midline of both the anterior and posterior walls. The ridges are known as the columns of the vagina; many rugae, or folds, extend from them to each side. The furrows between the rugae are more marked on the posterior wall and become especially pronounced before the birth of a child. The membrane undergoes little change during the menstrual cycle (except in its content of glycogen, a complex starchlike carbohydrate); this is in contradistinction to the situation in many mammals in which marked exfoliation (shedding of the surface cells) can occur. No glands are present in the vaginal lining, and mucus present has been secreted by the glands in the cervical canal of the uterus. The smooth muscle coat consists of an outer longitudinal layer and a less developed inner circular layer. The lower part of the vagina is surrounded by the bulbospongiosus muscle, a striped muscle attached to the perineal body.

The blood supply to the vagina is derived from several adjacent vessels, there being a vaginal artery from the internal iliac artery and also vaginal branches from the uterine, middle rectal, and internal pudendal arteries, all branches of the internal iliac artery. The nerve supply to the lower part of the vagina is from the pudendal nerve and from the inferior hypogastric and uterovaginal plexuses.

The Uterus

Uterine Structure

The uterus, or womb, is shaped like an inverted pear. It is a hollow, muscular organ with thick walls, and it has a glandular lining called the endometrium. In an adult the uterus is 7.5 cm (3 inches) long, 5 cm (2 inches) in width, and 2.5 cm (1 inch) thick, but it enlarges to four to five times this size in pregnancy. The narrower, lower end is called the cervix; this projects into the vagina. The cervix is made of fibrous connective tissue and is of a firmer consistency than the body of the uterus. The two fallopian tubes enter the uterus at opposite sides, near its top. The part of the uterus above the entrances of the tubes is called the fundus; the part below is termed the body. The body narrows toward the cervix, and a slight external constriction marks the juncture between the body and the cervix.

The uterus does not lie in line with the vagina but is usually turned forward (anteverted) to form approximately a right angle with it. The position of the uterus is affected by the amount of distension in the urinary bladder and in the rectum. Enlargement of the uterus in pregnancy causes it to rise up into the abdominal cavity, so that there is closer alignment with the vagina. The nonpregnant uterus also curves gently forward; it is said to be anteflexed. The uterus is supported and held in position by the other pelvic organs, by the muscular floor or diaphragm of the pelvis, by certain fibrous ligaments, and by folds of peritoneum. Among the supporting ligaments are two double-layered broad ligaments, each of which contains a fallopian tube along its upper free border and a round ligament, corresponding to the gubernaculum testis of the male, between its layers. Two ligaments—the cardinal (mackenrodt) ligaments—at each side of the cervix are also important in maintaining the position of the uterus.

The cavity of the uterus is remarkably small in comparison with the size of the organ. Except during pregnancy, the cavity is flattened, with front and rear walls touching, and is triangular. The triangle is inverted, with its base at the top, between the openings of the two fallopian tubes, and with its apex at the isthmus of the uterus, the opening into the cervix. The canal of the cervix is flattened from front to back and is somewhat larger in its middle part. It is traversed by two longitudinal ridges and has oblique folds stretching from each ridge in an arrangement like the branches of a tree. The cervical canal is 2.5 cm (about 1 inch) in length; its opening into the vagina is called the external os of the uterus. The external os is small, almost circular, and often depressed. After childbirth, the external os becomes bounded by lips in front and in back and is thus more slitlike. The cervical canal is lined by a mucous membrane containing numerous glands that secrete a clear, alkaline mucus. The upper part of this lining undergoes cyclical changes resembling, but not as marked as, those occurring in the body of the uterus. Numerous small cysts (nabothian cysts) are found in the cervical mucous membrane. It is from this region that cervical smears are taken in order to detect early changes indicative of cancer.

The uterus is composed of three layers of tissue. On the outside is a serous coat of peritoneum (a membrane exuding a fluid like blood minus its cells and the clotting factor fibrinogen), which partially covers the organ. In front it covers only the body of the cervix; behind it covers the body and the part of the cervix that is above the vagina and is prolonged onto the posterior vaginal wall; from there it is folded back to the rectum. At the side the peritoneal layers stretch from the margin of the uterus to each side wall of the pelvis, forming the two broad ligaments of the uterus.

The middle layer of tissue (myometrium) is muscular and comprises the greater part of the bulk of the organ. It is very firm and consists of densely packed, unstriped, smooth muscle fibres. Blood vessels, lymph vessels, and nerves are also present. The muscle is more or less arranged in three layers of fibres running in different directions. The outermost fibres are arranged longitudinally. Those of the middle layer run in all directions without any orderly arrangement; this layer is the thickest. The innermost fibres are longitudinal and circular in their arrangement.

The innermost layer of tissue in the uterus is the mucous membrane, or endometrium. It lines the uterine cavity as far as the isthmus of the uterus, where it becomes continuous with the lining of the cervical canal. The endometrium contains numerous uterine glands that open into the uterine cavity and are embedded in the cellular framework or stroma of the endometrium. Numerous blood vessels and lymphatic spaces are also present. The appearances of the endometrium vary considerably at the different stages in reproductive life. It begins to reach full development at puberty and thereafter exhibits dramatic changes during each menstrual cycle. It undergoes further changes before, during, and after pregnancy, during the menopause, and in old age. These changes are for the most part hormonally induced and controlled by the activity of the ovaries.

The Endometrium in the Menstrual Cycle

To understand the nature of the changes in the endometrium during each menstrual cycle it is usual to consider the endometrium to be composed of three layers. They blend imperceptibly but are functionally distinct: the inner two layers are shed at menstruation, and the outer or basal layer remains in position against the innermost layer of the myometrium. The three layers are called, respectively, the stratum compactum, the stratum spongiosum, and the stratum basale epidermidis. The stratum compactum is nearest to the uterine cavity and contains the lining cells and the necks of the uterine glands; its stroma is relatively dense. Superficial blood vessels lie beneath the lining cells. The stratum spongiosum is the large middle layer. It contains the main portions of uterine glands and accompanying blood vessels; the stromal cells are more loosely arranged and larger than in the stratum compactum. The stratum basale epidermidis lies against the uterine muscle; it contains blood vessels and the bases of the uterine glands. Its stroma remains relatively unaltered during the menstrual cycle.

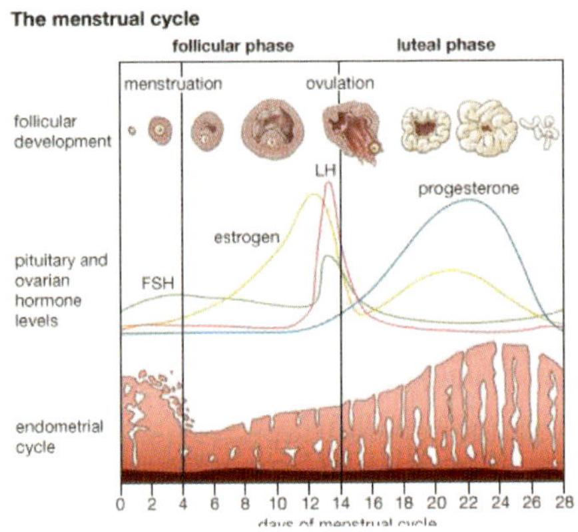

The menstrual cycle.

The menstrual cycle extends over a period of about 28 days (normal range 21–34 days), from the first day of one menstrual flow to the first day of the next. It reflects the cycle of changes occurring in the ovary, which is itself under the control of the anterior lobe of the pituitary gland. The menstrual cycle is divided into four phases: menstrual, postmenstrual, proliferative, and secretory.

The secretory phase reaches its climax about a week after ovulation. Ovulation occurs in midcycle, about 14 days before the onset of the next menstrual flow. The endometrium has been prepared and has been stimulated to a state of active secretion for the reception of a fertilized ovum. The stage has been set for the attachment of the blastocyst, derived from a fertilized ovum, to the endometrium and for its subsequent embedding. This process is called implantation; its success depends on the satisfactory preparation of the endometrium in both the proliferative and secretory phases. When implantation occurs, a hormone from certain cells of the blastocyst causes prolongation of the corpus luteum and its continued activity. This causes suppression of menstruation and results in the maintenance of the endometrium and its further stimulation by progesterone, with consequent increased thickening. The endometrium of early pregnancy is known as the decidua.

In a cycle in which fertilization of the ovum has not taken place, the secretory phase terminates in menstruation.

The endometrium needs to be in a certain state of preparedness before implantation can occur. When this stage has been passed, menstruation occurs. Repair then reestablishes an endometrium capable of being stimulated again to the critical stage when implantation can occur.

Blood Supply and Innervation

The uterus is supplied with blood by the two uterine arteries, which are branches of the internal iliac arteries, and by ovarian arteries, which connect with the ends of the uterine arteries and send branches to supply the uterus. The nerves to the uterus include the sympathetic nerve fibres, which produce contraction of uterine muscle and constriction of vessels, and parasympathetic (sacral) fibres, which inhibit muscle activity and cause dilation of blood vessels.

The Fallopian Tubes

The fallopian, or uterine, tubes carry ova from the ovaries to the cavity of the uterus. Each opens into the abdominal cavity near an ovary at one end and into the uterus at the other. Three sections of the tubes are distinguished: the funnel-shaped outer end, or infundibulum; the expanded and thin-walled intermediate portion, or ampulla; and the cordlike portion, the isthmus, that opens into the uterus. The infundibulum is fringed with irregular projections called fimbriae. One fimbria, somewhat larger than the others, is usually attached to the ovary. The opening into the abdomen is at the bottom of the infundibulum and is small. Fertilization of the ovum usually occurs in the ampulla of the tube. Normally the fertilized egg is transported to the uterus, but occasionally it may adhere to the tube and start developing as an ectopic pregnancy, or tubal pregnancy. The tube is unable to support this pregnancy, and the conceptus may be extruded through the abdominal opening or may cause rupture of the tube, with ensuing hemorrhage.

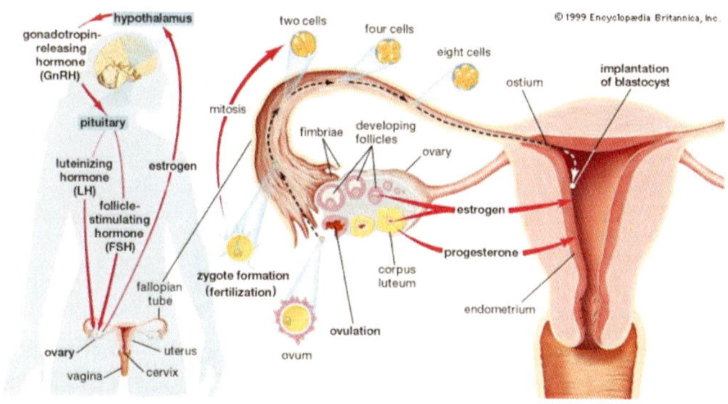

Major structures and hormones involved in the initiation of pregnancy.
At right, is the development of an egg cell (ovum) from follicle to embryo.

The fallopian tube is covered by peritoneum except on its border next to the broad ligament. There are inner circular and outer longitudinal layers of smooth muscle fibres continuous with those of the uterus. The inner lining has numerous longitudinal folds that are covered with ciliated columnar and secretory cells. Muscular contraction, movement of the hairlike cilia, and the passage of the watery secretions all probably assist in the transport of sperm to the ampulla and of a fertilized ovum toward the uterus.

The Ovaries

Ovarian Structure

The female gonads, or primary sex organs, corresponding to the testes in a male, are the two ovaries. Each is suspended by a mesentery, or fold of membrane, from the back layer of the broad ligament of the uterus. In a woman who has not been pregnant, the almond-shaped ovary lies in a vertical position against a depression, the ovarian fossa, on the side wall of the lesser pelvis. This relationship is altered during and after pregnancy. Each ovary is somewhat over 2.5 cm (1 inch) in length, 1.25 cm (0.5 inch) across, and slightly less in thickness, but the size varies much with age and with state of activity.

The mesentery of the ovary helps to keep it in position, and within this membrane lie the ovarian artery and vein, lymphatic vessels, and nerve fibres. The fallopian tube arches over the ovary and curves downward on its inner or medial surface.

Except at its hilum, the point where blood vessels and the nerve enter the ovary and where the mesentery is attached, the surface of the ovary is smooth and is covered by cubical cells. Beneath the surface, the substance of the ovary is divided into an outer portion, the cortex, and an inner portion, or medulla. The outermost part of the cortex, immediately beneath the outer covering, forms a thin connective tissue zone, the tunica albuginea. The rest of the cortex consists of stromal or framework cells, contained in a fine network of fibres, and also the follicles and corpora lutea.

The ovarian follicles, sometimes called graafian follicles, are rounded enclosures for the developing ova in the cortex near the surface of the ovary. At birth and in childhood they are present as numerous primary or undeveloped ovarian follicles. Each contains a primitive ovum, or oocyte, and each is covered by a single layer of flattened cells. As many as 700,000 primary follicles are contained in the two ovaries of a young female. Most of these degenerate before or after puberty.

Ovulation

During the onset of puberty and thereafter until menopause (except during pregnancy), there is a cyclic development of one or more follicles each month into a mature follicle. The covering layer of the primary follicle thickens and can be differentiated into an inner membrana granulosa and an outer vascularized theca interna. The cells of these layers (mostly the theca interna) produce estrogenic steroid hormones that exert their effects on the endometrium of the uterus and on other tissues. The maintenance and growth of the follicle to maturity is brought about by a follicle-stimulating hormone (FSH) from the anterior lobe of the pituitary gland. Another hormone, called luteinizing hormone (LH), from the anterior lobe, assists FSH to cause the maturing, now fluid-filled follicle to secrete estrogens. LH also causes a ripe follicle (1.0–1.5 cm [0.4–0.6 inch] in diameter) to rupture, causing the liberation of the oocyte into the peritoneal cavity and thence into the fallopian tube. This liberation of the oocyte is called ovulation; it occurs at about the midpoint of the reproductive cycle, on the 13th or 14th day of a 28-day cycle as measured from the first day of the menstrual flow.

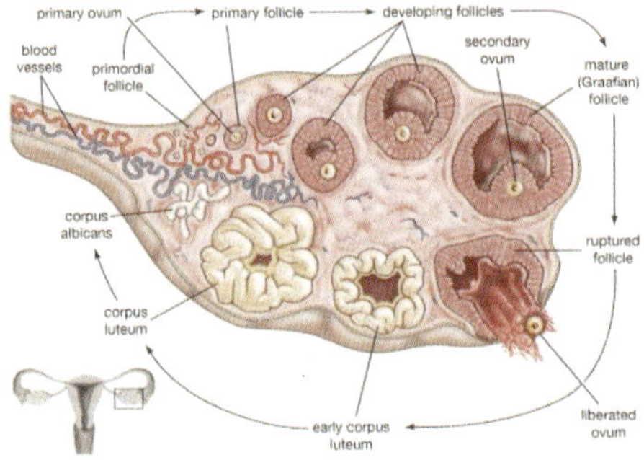

The steps of ovulation, beginning with a dormant primordial follicle that grows and matures and is eventually released from the ovary into the fallopian tube.

After ovulation the ruptured follicle collapses because of loss of its follicular fluid and rapidly becomes transformed into a soft, well-vascularized glandular structure known as the corpus luteum("yellow body"). The corpus luteum develops rapidly, becomes vascularized after about four days, and is fully established by nine days. The gland produces the steroid hormone progesterone and some estrogens. Its activity is both stimulated and maintained by luteinizing hormone. Progesterone stimulates glandular proliferation and secretion in an endometrium primed by estrogens.

While the ovarian follicle matures, the primary oocyte divides into a secondary oocyte and a small rudimentary ovum called the first polar body. This occurs at about the time when the follicle develops its cavity; the oocyte also gains a translucent acellular covering, or envelope, the zona pellucida. The secondary oocyte is liberated at ovulation; it is 120–140 micrometres in diameter and is surrounded by the zona pellucida and a few layers of cells known as the corona radiata. The final maturation of the oocyte, with the formation of the rudimentary ovum called the second polar body, occurs at the time of fertilization.

If fertilization does not occur, then the life of the corpus luteum is limited to about 14 days. Degeneration of the gland starts toward the end of this period, and menstruation occurs. The corpus luteum shrinks, fibrous tissue is formed, and it is converted into a scarlike structure called a corpus albicans, which persists for a few months.

Should fertilization occur and be followed by implantation of the blastocyst, hormones (particularly human chorionic gonadotropin) are produced by cells of the blastocyst to prolong the life of the corpus luteum. It persists in an active state for at least the first two months of pregnancy, until the placental tissue has taken over its hormone-producing function. The corpus luteum of pregnancy then also retrogresses, becoming a fibrous scar by the time of parturition.

Blood Supply and Innervation

The ovarian arteries arise from the front of the aorta in a manner similar to the testicular arteries, but at the brim of the lesser pelvis they turn down into the pelvic cavity. Passing in the suspensory ligament of the ovary, each artery reaches the broad ligament below the fallopian tube and then passes into the mesovarium to divide into branches distributed to the ovary. One branch continues in the broad ligament to anastomose with the uterine artery. The ovarian veins emerge from each ovary as a network that eventually becomes a single vein; the terminations are similar to those of the testicular veins. The nerves are derived from the ovarian nerve network on the ovarian artery.

Process of Human Reproduction

The process of reproduction in humans usually relies on sexual intercourse between a male and a female, although there are exceptions to this. Unlike many animals, humans mate throughout the year. Humans have sexual intercourse when sexual reproduction is not possible for reasons such as the use of birth control or female menopause. Practices and behaviors surrounding human reproduction vary widely across cultures, but in every case, it involves sperm, an ovum (egg), a uterus and a baby.

Gamete Production

The process of reproduction in humans begins with meiosis. In human meiosis, diploid cells with the usual 46 chromosomes divide into four haploid daughter cells, each containing 23 chromosomes. Each of these daughter cells is called a gamete. In males, this meiotic process is called spermatogenesis, and the daughter cells are sperm. In females, the process is called oogenesis, and the daughter cells are called ova. Males begin spermatogenesis at puberty and continue throughout life. Healthy young adult males produce hundreds of millions of sperm each day. This number begins to decline by their mid-20s.

Unlike males, females begin to produce gametes before they even are even born. By the fifth month in the womb, female fetuses have begun oogenesis, but the process pauses after a phase called prophase I, suspending the ova in the primary oocyte stage until puberty. 99.9 percent of a female's ova remain in the primary oocyte stage until they are eventually absorbed by the body. Millions are absorbed by the time a fetus is born, and by puberty, only 400,000 remain. For every ovulation, about 2,000 more ova are absorbed.

Sexual Intercourse

The four stages of the human sexual response cycle occur during partnered sex with people of any gender, as well as other sexual stimulation activities. The first stage is excitement, the beginning of arousal, in which blood flow increases and causes engorgement in the genitals and nipples, accompanied by an increase in heart rate, breathing rate, muscle tone and blood pressure. Next is the plateau stage, which is brief, and which involves an increase in arousal.

The third phase is orgasm, which involves waves of muscle spasms and pleasure that last several seconds. During this phase, the uterus has several contractions, and the penis has contractions at its base, causing semen, the fluid containing sperm, to ejaculate into the vagina. The last stage is resolution, during which the body relaxes to its original state.

1. Fertilization: a sperm and an egg form a zygote

During sexual intercourse, some sperm ejaculated from the male penis swim up through the female vagina and uterus toward an oocyte (egg cell) floating in one of the uterine tubes. The sperm and the egg are gametes. They each contain half the genetic information necessary for reproduction. When a sperm cell penetrates and fertilizes an egg, that genetic information combines. The 23 chromosomes from the sperm pair with 23 chromosomes in the egg, forming a 46-chromosome cell called a zygote. The zygote starts to divide and multiply. As it travels toward the uterus it divides to become a blastocyst, which will burrow into the uterine wall.

2. The zygote becomes an embryo: development prior to and during implantation

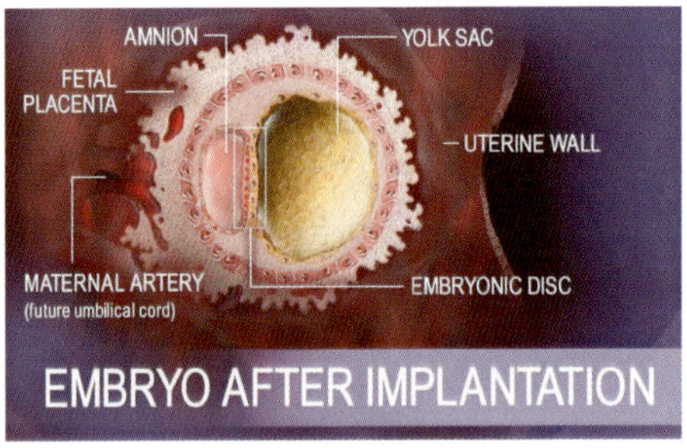

A fertilized egg, or zygote, takes about five days to reach the uterus from the uterine tube. As it moves, the zygote divides and develops into a blastocyst, with an inner mass of cells and a protective outer

ring. The blastocyst attaches to the wall of the uterus and gradually implants itself into the uterine lining. During implantation, its cells differentiate further. At day 15 after conception, the cells that will form the embryo become an embryonic disc. Other cells begin to form support structures. The yolk sac, on one side of the disc, will become part of the digestive tract. On the other side, the amnion fills with fluid and will surround the embryo as it develops. Other cell groups initiate the placenta and umbilical cord, which will bring in nutrients and eliminate waste.

3. In eight weeks, the embryo develops; by the end of week 10 it becomes a fetus

Fifteen days after conception marks the beginning of the embryonic period. The embryo contains a flat embryonic disc that now differentiates into three layers: the endoderm, the mesoderm, and the ectoderm. All organs of the human body derive from these three tissues. They begin to curve and fold and to form an oblong body. By week 4, the embryo has a distinct head and tail and a beating heart. Over the next six weeks, limbs, eyes, brain regions, and vertebrae form. Primitive versions of all body systems appear. By the end of week 10, the embryo is a fetus. (Pregnancy is often measured in terms of gestational age—age of the fetus starting with the first day of a woman's last menstrual period—and embryonic or fetal age—actual age of the growing fetus. We are referring to the gestational age of the fetus.)

4. Fetal development ends with labor, a three-stage process

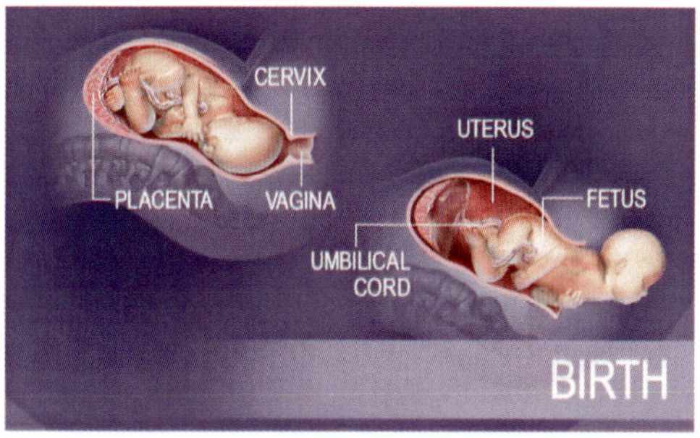

From week 10 of pregnancy, the fetus grows inside the uterus, fueled by nutrient-rich blood supplied by the umbilical cord. The placenta provides oxygen and nutrients to the fetus and removes waste products from the fetus' blood. Bones, muscles, skin, and connective tissues form. Body systems develop. Limbs and facial features take shape. Around week 36 (usually), the process of labor begins. In the first stage, dilation, hormones stimulate downward contractions of the uterine walls. The contractions push the head of the fetus against the cervix at the lower end of the uterus. The cervix dilates. In the second stage, expulsion, powerful contractions push the head and the rest of the body through the dilated cervix, and out through the vagina and the vulva. The baby is born. Further contractions expel the placenta to complete the placental stage.

5. Humans must develop male or female gonads and genitalia to be capable of reproduction

Reproductive structures begin to form in the embryonic stage. By week 6, gonads and genitalia are present but undifferentiated. Whether they become male or female is determined by one chromosome delivered by the sperm. This pair contains an X sex chromosome from the female egg and either an X or a Y sex chromosome from the male sperm. If the chromosome pair is XY, the gonads develop into testes starting in week 7. If the chromosome pair is XX, the gonads become ovaries starting in week 8. Testes secrete testosterone, forming male genitalia around week 10. Without testosterone, female genitalia form. All reproductive structures are in place at birth or shortly after. At puberty, an increase in sex hormones will grow them to their adult size and reproductive capability.

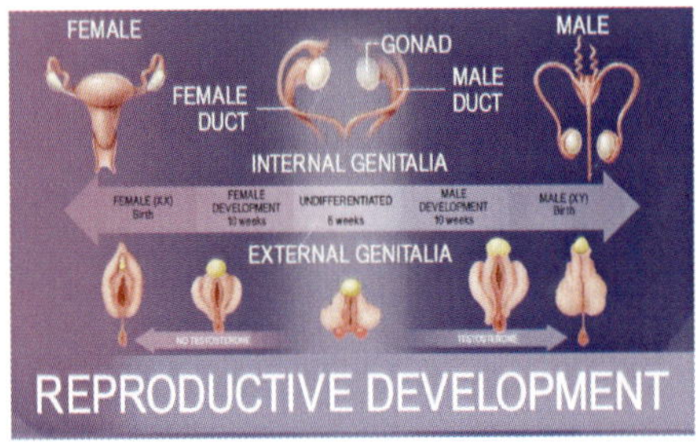

Hormonal Control of Reproduction

The human male and female reproductive cycles are controlled by the interaction of hormones from the hypothalamus and anterior pituitary with hormones from reproductive tissues and organs. In both sexes, the hypothalamus monitors and causes the release of hormones from the anterior pituitary gland. When the reproductive hormone is required, the hypothalamus sends a gonadotropin-releasing hormone (GnRH) to the anterior pituitary. This causes the release of follicle stimulating hormone (FSH) and luteinizing hormone (LH) from the anterior pituitary into the blood. Although these hormones are named after their functions in female reproduction, they are produced in both sexes and play important roles in controlling reproduction. Other hormones have specific functions in the male and female reproductive systems.

Male Hormones

At the onset of puberty, the hypothalamus causes the release of FSH and LH into the male system for the first time. FSH enters the testes and stimulates the sertoli cells located in the walls of the seminiferous tubules to begin promoting spermatogenesis (figure). LH also enters the testes and stimulates the interstitial cells of Leydig, located in between the walls of the seminiferous tubules, to make and release testosterone into the testes and the blood.

Testosterone stimulates spermatogenesis. This hormone is also responsible for the secondary sexual characteristics that develop in the male during adolescence. The secondary sex characteristics in males include a deepening of the voice, the growth of facial, axillary, and pubic hair, an increase in muscle bulk, and the beginnings of the sex drive.

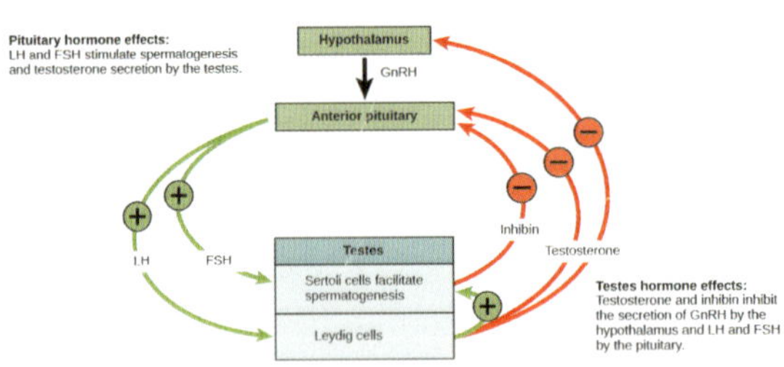

Figure: Hormones control sperm production in a negative feedback system.

A negative feedback system occurs in the male with rising levels of testosterone acting on the hypothalamus and anterior pituitary to inhibit the release of GnRH, FSH, and LH. In addition, the sertoli cells produce the hormone inhibin, which is released into the blood when the sperm count is too high. This inhibits the release of GnRH and FSH, which will cause spermatogenesis to slow down. If the sperm count reaches a low of 20 million/mL, the Sertoli cells cease the release of inhibin, and the sperm count increases.

Female Hormones

The control of reproduction in females is more complex. The female reproductive cycle is divided into the ovarian cycle and the menstrual cycle. The ovarian cycle governs the preparation of endocrine tissues and release of eggs, while the menstrual cycle governs the preparation and maintenance of the uterine lining. These cycles are coordinated over a 22–32 day cycle, with an average length of 28 days.

As with the male, the GnRH from the hypothalamus causes the release of the hormones FSH and LH from the anterior pituitary. In addition, estrogen and progesterone are released from the developing follicles. As with testosterone in males, estrogen is responsible for the secondary sexual characteristics of females. These include breast development, flaring of the hips, and a shorter period for bone growth.

References

- Sexual-reproduction-in-plants: lumenlearning.com, Retrieved 14 April 2018
- Human-reproductive-system, science: britannica.com, Retrieved 25 May 2018
- Process-reproduction-humans-5406051: sciencing.com, Retrieved 18 July 2018
- Reproductive-process, reproductive: visiblebody.com, Retrieved 12 May 2018
- 13-3-human-reproduction, biology: opentextbc.ca, Retrieved 25 April 2018

6

Population Ecology

Population ecology is a sub-discipline of ecology. The study of the interaction of species population with the environment and the change in population sizes over time and space is under the scope of population ecology. All the vital topics of this field like population density, population age structure, population growth, population dispersion and limits to population growth have been carefully examined in this chapter.

Population ecology is the study of the processes that affect the distribution and abundance of animal and plant populations. A population is a subset of individuals of one species that occupies a particular geographic area and, in sexually reproducing species, interbreeds. The geographic boundaries of a population are easy to establish for some species but more difficult for others. For example, plants or animals occupying islands have a geographic range defined by the perimeter of the island. In contrast, some species are dispersed across vast expanses, and the boundaries of local populations are more difficult to determine. A continuum exists from closed populations that are geographically isolated from, and lack exchange with, other populations of the same species to open populations that show varying degrees of connectedness.

Genetic Variation within Local Populations

In sexually reproducing species, each local population contains a distinct combination of genes. As a result, a species is a collection of populations that differ genetically from one another to a greater or lesser degree. These genetic differences manifest themselves as differences among populations in morphology, physiology, behaviour, and life histories; in other words, genetic characteristics (genotype) affect expressed, or observed, characteristics (phenotype). Natural selection initially operates on an individual organismal phenotypic level, favouring or discriminating against individuals based on their expressed characteristics. The gene pool (total aggregate of genes in a population at a certain time) is affected as organisms with phenotypes that are compatible with the environment are more likely to survive for longer periods, during which time they can reproduce more often and pass on more of their genes.

The amount of genetic variation within local populations varies tremendously, and much of the discipline of conservation biology is concerned with maintaining genetic diversity within and among populations of plants and animals. Some small isolated populations of asexual species often have little genetic variation among individuals, whereas large sexual populations often have great variation. Two major factors are responsible for this variety: mode of reproduction and population size.

Effects of Mode of Reproduction: Sexual and Asexual

In sexual populations, genes are recombined in each generation, and new genotypes may result. Offspring in most sexual species inherit half their genes from their mother and half from their father, and their genetic makeup is therefore different from either parent or any other individual in the population. In both sexually and asexually reproducing species, mutations are the single most important source of genetic variation. New favourable mutations that initially appear in separate individuals can be recombined in many ways over time within a sexual population.

In contrast, the offspring of an asexual individual are genetically identical to their parent. The only source of new gene combinations in asexual populations is mutation. Asexual populations accumulate genetic variation only at the rate at which their genes mutate. Favourable mutations arising in different asexual individuals have no way of recombining and eventually appearing together in any one individual, as they do in sexual populations.

Effects of Population Size

Over long periods of time, genetic variation is more easily sustained in large populations than in small populations. Through the effects of random genetic drift, a genetic trait can be lost from a small population relatively quickly. For example, many populations have two or more forms of a gene, which are called alleles. Depending on which allele an individual has inherited, a certain phenotype will be produced. If populations remain small for many generations, they may lose all but one form of each gene by chance alone.

This loss of alleles happens from sampling error. As individuals mate, they exchange genes. Imagine that initially half of the population has one form of a particular gene, and the other half of the population has another form of the gene. By chance, in a small population the exchange of genes could result in all individuals of the next generation having the same allele. The only way for this population to contain a variation of this gene again is through mutation of the gene or immigration of individuals from another population.

Minimizing the loss of genetic variation in small populations is one of the major problems faced by conservation biologists. Environments constantly change, and natural selection continually sorts through the genetic variation found within each population, favouring those individuals with phenotypes best suited for the current environment. Natural selection, therefore, continually works to reduce genetic variation within populations, but populations risk extinction without the genetic variation that allows populations to respond evolutionarily to changes in the physical environment, diseases, predators, and competitors.

There are a number of characteristics of populations that help ecologist and other scientists to monitor and manage wild populations. Population density, abundance, distribution, age structure and sex ratio are important characteristics which can be monitored for good population management.

Density

Population density refers to the number of individuals in a given space. For instance, density may be measured as the number of trees per hectare or the number of wolves per square mile. The abundance of a population is different to the density as it does not deal with an area or volume but focuses entirely on the total population size in a particular ecosystem.

Abundance can be difficult to measure and there a number of methods used to estimate a population's abundance if a complete count is not possible. Mark/recapture methods are common among large animal species where a number of individuals are caught, tagged and released; then an effort is made to recapture individuals from the same population and the percentage of caught individuals that are tagged is used to give an estimate of the population size for a given area.

Distribution of a Population

The distribution of a population tells us how large the home range is that individuals from a particular population can be found. Generally, populations are not distributed evenly and so density and distribution are often studied together.

The way in which a population is distributed often gives an insight into the relationships between individuals. If individuals are spread evenly across an area it is often due to negative relationships that cause each individual to be as far away as possible from others. Contrary to an even spread, individuals of a population may be attracted to each other and live in clumps causing a very uneven distribution of the population across the landscape.

Age Structure

The age structure of a population can be helpful in determining the future growth rate of a population. Individuals can be categorized into three groups – pre-reproductive, reproductive and post-reproductive. Populations with large pre-reproductive and reproductive groups are likely to experience growth in the near future.

If large proportions of populations are post-reproductive it means recruitment of new individuals is proportionally low for the population. This is likely to lead to either zero growth or a decrease in the population's size.

Age structure is very important in the management of fisheries and wild populations. In management schemes it is important to maintain significant levels of reproductive individuals and in harvested stocks knowledge of the populations age structure can be used to influence which individuals are important to keep in the population.

Sex Ratio

The sex ratio of a population is simply the number of males relative to females. Sex ratio is often influenced with time and the life stage of individuals. The sex ratio at fertilization is usually around 1:1 but this has often changed by the time individuals reach a sexually active age, where females regularly have an increased abundance. At the post-reproductive stage, female abundance is also often higher than that of their male counter-parts. These are, however, only trends and differ from species to species.

Meta Populations

Meta-populations are groups of separated populations which interact with each other at some level. There is a range of different functional structuring that meta-populations can be built upon. The structure of meta-populations depends on the level of isolation/connectivity between meta-populations and the size structure of each sub-population. The sub-population being a specific group of a meta-population, the meta-population is therefore made up from many sub-populations.

Population Ecology Relative to Other Ecological Disciplines

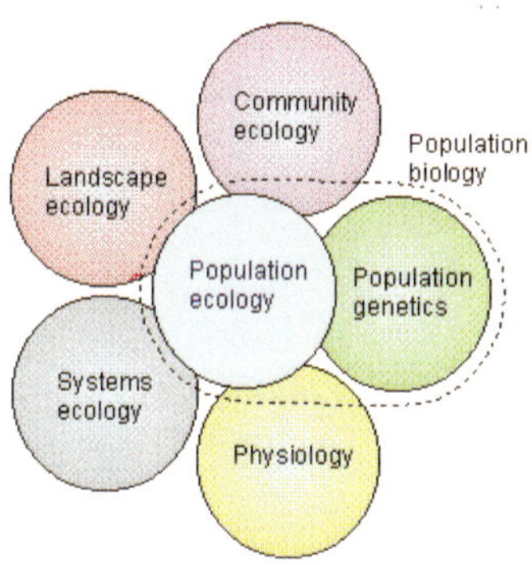

- Population ecology is the branch of ecology that studies the structure and dynamics of populations.
- Physiology studies individual characteristics and individual processes. These are use as a basis for prediction of processes at the population level.
- Community ecology studies the structure and dynamics of animal and plant communities. Population ecology provides modeling tools that can be used for predicting community structure and dynamics.
- Population genetics studies gene frequencies and microevolution in populations. Selective advantages depend on the success of organisms in their survival, reproduction and competition. And these processes are studied in population ecology. Population ecology and population genetics are often considered together and called "population biology". Evolutionary ecology is one of the major topics in population biology.
- Systems ecology is a relatively new ecological discipline which studies interaction of human population with environment. One of the major concepts are optimization of ecosystem exploitation and sustainable ecosystem management.

- Landscape ecology is also a relatively new area in ecology. It studies regional large-scale ecosystems with the aid of computer-based geographic information systems. Population dynamics can be studied at the landscape level, and this is the link between landscape- and population ecology.

Population Density

Population density is a statistic that tells us how many people live in a certain area.

In general, pop. density is reported in square miles or kilometers, and may or may not include bodies of water. This type of measurement is called arithmetic density, and is reported as the total number of people per land area. For example, as of 2014, the Nile Delta has a pop. density of 1000 people per square kilometer. In other words, an average of 1000 people are living in an area 1 x 1 km.

Formula

The formula for population density is:

number of people / square miles(or kilometers) of land.

The definition differs slightly according to which organization or entity is reporting the figure. For example:

- The World Bank defines population density as "midyear population divided by land area in square kilometers." World Bank figures include all people living in a land area. This includes illegal aliens but excludes refugees (who are considered residents of their country of origin) and other temporary residents. "Land area" excludes major rivers and lakes.
- The University of Florida calculates population density by first excluding blocks of land where there are no people living (like parks, wetlands and forests). The university states that including this area (38% of Florida is completely uninhabited) can lead to misleading statistics about true population densities of where people actually live.

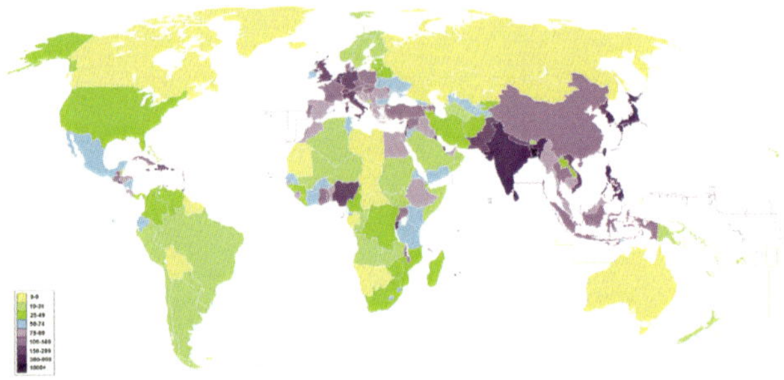

World Population Density map.

Apart from arithmetic density, less common types of pop. density measurement include:

- Residential density: number of people living in urban areas / area of residential land.
- Urban density: number of people living in an urban area / area of urban land.

- Agricultural density (also called physiological density): number of people in a rural area/area of arable land.

Calculation of Population Density

Choose the area that you would like to calculate the population density for. The smaller your area gets, the more accurately your calculations will describe that particular area. Larger areas give you a much worse estimate of the population density at a given location. For example, if you calculated the population density of New York as a whole state, you might get a pretty average population density. However, if you calculated the population density of an area in upstate NY and an area in Manhattan, you would come up with two very different numbers, each of which would describe their area better.

Find the area of the place that you want to calculate the population density for. This area can be expressed in square miles, acres, or square meters, depending on your source. Just make sure that you keep the units constant through your whole calculation.

Find the population for your selected area.

Make a fraction with the correct units for population density. Scientists and sociologists usually express population density as a number of people per unit of area. Therefore, to create a correct fractional expression, you would need to have the number of people in the numerator (on the top of the fraction) and the number of square miles/meters/acres in the denominator.

For example, a population of 8,341 people living in an area of 25 square miles would be expressed as a fraction: 8341/25.

Reduce the fraction to a unit area so that the final result is the number of people in one acre or square mile or square meter, for example. Another way to think of this is to carry out the division represented by the fraction. If you wanted to calculate the population density of a town that is 25 square miles and has a population of 8,341 people, you would divide 8,341 by 25 to get 333.64 people per square mile.

Round your answer from Step 5 up or down to a whole number, if necessary. The population density is simply an average, so you can end up with a decimal; however, the real population is composed of whole people, so rounding will express your answer in whole people as well.

Population Age Structure

The age structure of a population is the distribution of people among various ages. It is a useful tool for social scientists, public health and health care experts, policy analysts, and policy-makers because it illustrates population trends like rates of births and deaths. These are important to understand because they have a host of social and economic implications in society, like understanding the resources that must be allocated for childcare, schooling, and healthcare, and the familial and greater social implications of whether there are more children or elderly in society. In graphic form, age structure is portrayed as an age pyramid that shows the youngest age cohort at the bottom, with each additional layer showing the next oldest cohort. Typically males are indicated on the left and females on the right.

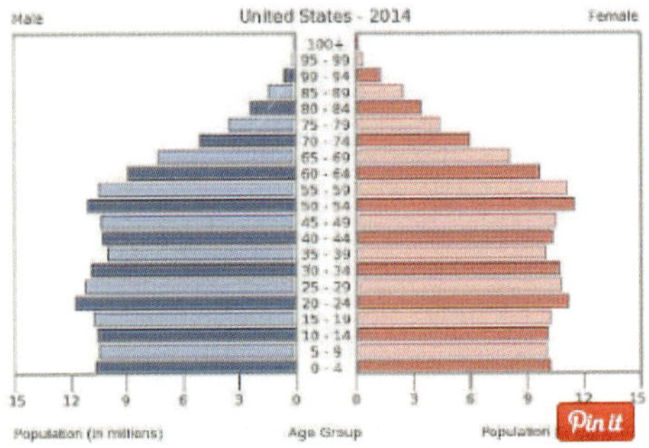

This age pyramid illustrates the age structure of the population of the United States in 2014.

Concepts and Implications

Both age structure and age pyramids can take a variety of forms, depending on the birth and death trends within the population, as well as a host of other social factors. They can be stable, meaning that patterns of birth and death are unchanging over time; stationary, which signals both low birth and death rates (they slope gently inward and have a rounded top); expansive, which slope dramatically inward and upward from the base, indicate that a population has both high birth and death rates; or constrictive, which signal low birth and death rates, and expand outward from the base before sloping inward to achieve a rounded peak at the top.

The current U.S. age structure and pyramid, shown above, is a constrictive model, which is typical of developed countries where family planning practices are common and access to birth control is (ideally) easy, and where advanced medicine and treatments are commonly available through accessible and affordable healthcare (again, ideally). This pyramid shows us that the birth rate has slowed in recent years because we can see that there are more teens and young adults in the U.S. today than there are young children (the birth rate is lower today than it was in the past). That the pyramid moves stably upward through age 59, then only gradually shrinks inward through age 69, and only gets really narrow after age 79 shows us that people are living long lives, which means that the death rate is low. Advances in medicine and elder care over the years have produced this effect in developed countries.

The U.S. age pyramid also shows us how birth rates have shifted over the years. The millennial generation is now the largest in the U.S., but it is not so much larger than Generation X and the Baby Boomer generation, who are now in their 50s and 60s. This means that while birth rates have increased a bit over time, more recently they have declined. However, the death rate has declined considerably, which is why the pyramid looks the way it does.

Many social scientists and health care experts are concerned about current population trends in the U.S. because this large population of teens, adults, and older adults are likely to have long lives, which will put a strain on an already underfunded social security system.

It is implications like this that make the age structure an important tool for social scientists and policy-makers.

Population Growth

Population growth refers to change in the size of a population—which can be either positive or negative—over time, depending on the balance of births and deaths. If there are many deaths, the world's population will grow very slowly or can even decline. Population growth is measured in both absolute and relative terms. Absolute growth is the difference in numbers between a population over time; for example, in 1950 the world's population was 4 billion, and in 2000 it was 6 billion, a growth of 2 billion. Relative growth is usually expressed as a rate or a percentage; for example, in 2000 the rate of global population growth was 1.4 percent (or 14 per 1,000). For every 1,000 people in the world, 14 more are being added per year.

For the world as a whole, population grows to the extent that the number or rate of births exceeds the number or rate of deaths. The difference between these numbers (or rates) is termed "natural increase" (or "natural decrease" if deaths exceed births). For example, in 2000 there were 22 births per 1,000 population (the number of births per 1,000 population is termed the "crude birth rate") and 9 deaths per 1,000 population (the "crude death rate"). This difference accounts for the 2000 population growth rate of 14 per 1,000, which is also the rate of natural increase. In absolute numbers, this means that approximately 78 million people—or about the population of the Philippines—are added to the world each year. For countries, regions, states, and so on, population growth results from a combination of natural increase and migration flows. The rate of natural increase is equivalent to the rate of population growth only for the world as a whole and for any smaller geographical units that experience no migration.

Populations can grow at an exponential rate, just as compound interest accumulates in a bank account. One way to assess the growth potential of a population is to calculate its doubling time—the number of years it will take for a population to double in size, assuming the current rate of population growth remains unchanged. This is done by applying the "rule of seventy"; that is, seventy divided by the current population growth rate (in percent per year). The 1.4 percent global population growth rate in 2000 translates into a doubling time (if the growth rate remains constant) of fifty-one years.

The world's population grew very slowly until about 1750. There was a long period of stationary growth (no growth) until 1000 B.C.E. , when the world's population was approximately 300 million; this was followed by a period of slow growth from 1000 B.C.E. to approximately 1750, at which time global population was an estimated 800 million. Until this time, the world's population was kept in check by high death rates, which were due to the combined effects of plagues, famines, unsanitary living conditions, and general poverty. After 1750, the world's population grew substantially; by 1950 it had tripled to around 2.5 billion. In this 200-year period, the doubling time was 122 years. Growth from 1950 to 1985 was even more dramatic; by 1985, the human population was 5 billion. World population had doubled in thirty-five years. By 2000 global population was 6 billion and is projected to be 9 billion in 2050.

Population growth did not become exponential until around 1750. Before that, high mortality counter balanced the high fertility needed by agrarian parents. Death rates were high and life expectancy was low; life expectancy at birth was in the range of twenty to forty years (most likely around thirty years) until the middle of the eighteenth century. This high mortality was a function of several factors, including poor nutrition, which led directly to deaths through starvation and indirectly through increasing susceptibility to disease; epidemics; and, quite possibly, infanticide and geronticide, especially during times of food shortage.

Starting in the middle of the eighteenth century, the mortality rate began to decline in the West, the first place in the world where the natural balance between births and deaths was altered by humans. This decline in deaths occurred not because of major medical breakthroughs (e.g., penicillin was first used only in the 1940s) but rather because of improvements in food availability, housing, water cleanliness, personal hygiene, and public sanitation. Later, in the twentieth century, medical advances, particularly vaccinations against infectious diseases, accelerated mortality decline.

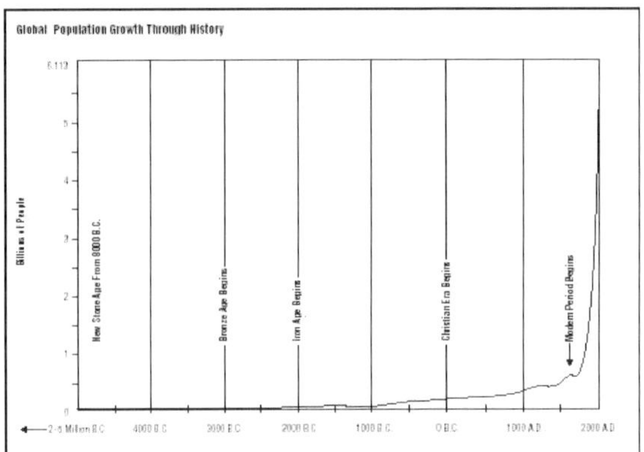

SOURCE: Yaukey, David and Douglas L. Anderton. *Demography: The Study of Human Population*. Prospect Heights, IL: Waveland Press, 2001.

Western mortality decline was relatively slow, paralleling socioeconomic development, and it occurred in a global context in which European population "surplus" (arising from gaps between lowering mortality and more slowly lowering fertility) was able to migrate to new areas (e.g., the United States, Canada, and Australia) that were very sparsely populated by Aboriginal peoples (whose numbers were reduced even more by contagious diseases brought by Europeans).

Mortality decline in less developed countries followed a different path. First, mortality decreases did not begin until around 1950, much later than in the West. Second, in many less developed countries, substantial mortality reductions occurred in a short period of time. A classic example is Ceylon (now Sri Lanka), where the death rate was halved in less than a decade, starting in the early 1950s. (In the West, a comparable reduction typically took around one century.) In these less developed countries, mortality decreases were not matched by fertility decreases, where they produce population growth rates much greater than those experienced in the West. So the demographic transition that took two centuries to unfold in the West occurred (or is occurring) within the span of a single life. Third, mortality decline did not parallel economic development. Rather, the impetus behind third world mortality reductions originated, for the most part, in factors external to the society. For example, the speedy mortality decline in Ceylon was due to the importation of American technology (pesticides and the airplanes for spraying them) that killed the mosquitoes that were responsible for malaria, the leading cause of death. During the cold war, it was not uncommon for the United States to provide non aligned countries with such assistance in the hopes of wooing them away from the Soviet Union and a communist development model.

As a result, the world witnessed unprecedented rapid population growth between 1950 and 1985, owing, in large part, to third world increases.

World Population and Population Growth Rate, 1950–2050

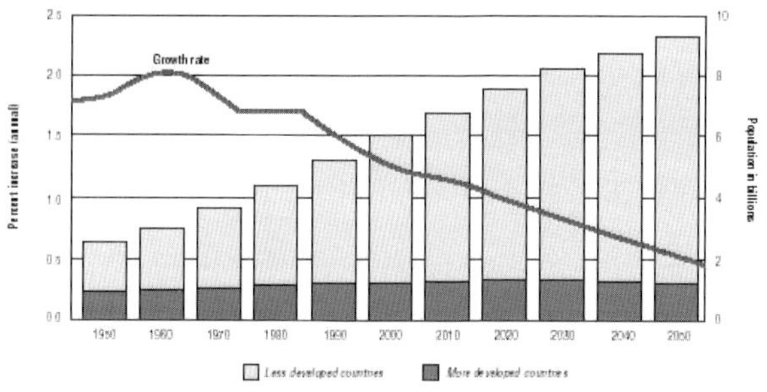

SOURCE: United Nations. *World Population Prospects: The 2000 Revision.* New York: United Nations, 2001.

Further, the phenomenal increase in human numbers over the past 250 years is largely the consequence of mortality declines—not fertility increases. The first deaths to be reduced were those due to infectious diseases, the victims of which were most often children. The old killers of the past were to be replaced by chronic and degenerative diseases; the primary victims shifted from the young to the old.

Population Growth 1950–2050

The rate of global population growth has declined significantly from its 1970s highs. Current estimates anticipate a continued decline to about 0.5 percent in 2050. This corresponds to a doubling time of 140 years, a rate that has fostered concern about how the world will cope with 18 billion people in 2190.

It is in the less developed countries that the continued growth in population will occur in the twenty-first century. Even though mortality is much higher in less developed countries (e.g., life expectancy at birth in 2000 was 75 years in the more developed countries and 62 to 64 years in the less developed countries), fertility remains even higher, thus accounting for relatively high growth in the third world. However, projections are not guarantees. Population may grow more slowly if, optimistically, fertility declines more quickly than experts expect (e.g., between just 1965 and 1987 the average number of children born to Thai women dropped from 6.3 to 2.2) or, pessimistically, if mortality increases, especially in light of the persistence of HIV/AIDS pandemic and other communicable diseases.

Theories of Population Growth

While theories about population growth first appeared in ancient Greece, the English clergyman and economist Thomas Malthus is considered to be the pioneering theorist of the modern age. Malthus formulated a "principle of population" that held that unchecked population grows more quickly than the means of subsistence (food and resources) to sustain it. Population will be controlled either by preventive checks (lowering the number of births, particularly by postponement of marriage age) or by positive checks (increasing deaths as a result of famines, plagues, natural disasters, war). Given a morally based preference for preventive checks, later followers of Malthus (neo-Malthusians) have supported family planning and contraception even though Malthus himself felt that contraception was unacceptable. Other neo-Malthusians have focused upon the claimed negative effects of rapid population growth: war, violence, and environmental degradation.

Karl Marx's views on population were directly opposed to those of Malthus. Marx disagreed with the Malthusian idea of a universal principle of population that applied to all societies. For Marx, population growth depended upon the economic base of society. Thus, capitalist society is characterized by its own population principle, which Marx termed the "law of relative population surplus." He argued that capitalism creates overpopulation (i.e., a surplus of people relative to jobs), leading to increased unemployment, cheap labor, and poverty. Also, capitalism requires unemployment in order to ensure a docile, low-paid class of laborers. Marx envisioned that overpopulation would not occur in post capitalist, communist society.

In the middle of the twentieth century, demographic transition theory became the dominant theory of population growth. Based on observed trends in Western European societies, it argues that populations go through three stages in their transition to a modern pattern. Stage One (pre transition) is characterized by low or no growth, and high fertility is counterbalanced by high mortality. In Stage Two (the stage of transition), mortality rates begin to decline, and the population grows at a rapid pace. By the end of this stage, fertility has begun to decline as well. However, because mortality decline had a head start, the death rate remains lower than the birth rate, and the population continues to experience a high rate of growth. In Stage Three (post transition), the movement to low fertility and mortality rates is complete, producing once again a no-growth situation. The theory of demographic transition explains these three stages in terms of economic development, namely industrialization and urbanization. Since about 1980, demographic transition theory has been criticized on a number of grounds, including its assumption that the demographic experience of non-Western societies will inevitably follow that of the West; its failure to consider cultural variables; and its hypothesized relationship between population growth and economic development. Indeed, all three theories above contain assumptions about population growth and economic development; however, there is mounting evidence that this relationship is complex and varies from context to context. As the twenty-first century begins, the attempt to erect a general theory of population growth has been abandoned, signaling for some an alarming trend in population studies.

Population Dispersion

Population dispersion is the observation of where individuals are found in a habitat. How individuals "disperse" themselves. There are three main types of dispersion: clumped, uniform and random.

Clumped Dispersion

Is the tendency for populations to be found in tight clusters, dispersed across a large landscape. In between these population hubs, very few to no individuals are usually found. This sort of a dispersion can be caused by a number of factors. Some species cluster together for protection, while others group around natural resources necessary to their survival.

Uniform Dispersion

Is the tendency for populations to be found evenly distributed about their habitat. This is generally caused by a species ability to survive anywhere in their habitat - they use the resources found immediately around them, and spread out as to use all of the available resources.

Random Dispersion

Is the tendency for populations to be found randomly about their habitat. In immobile species, this is usually caused by their ability to live anywhere in a given habitat, except, they are limited to growing wherever they are first set root (which is usually caused randomly, from spores drifting in the wind to seeds falling and tumbling on the ground). In motile populations, individuals are able to move about their habitat, so that at any given instance, they can be found anywhere about their environment.

Process of Population Sampling

Using Quadrats to Calculate Populations

Using quadrats is a very simple way of accurately calculating immobile populations. The procedure is very simple - the number of individuals in a measured area (the area being a tiny fraction of the habitat of the species being counted) is counted, then averaged to get a population density. The total area of the species habitat is also figured out, then by multiplying the area of the habitat by the average density of individuals in an area, the total population can be derived.

For Example:

Counting the number of blades of grass in a 200 square meter field would be extremely difficult and time consuming. So instead, several samples are taken from quadrats taken from around the field. If the average density of blades of grass found in these quadrants is 10,000 blades of grass per meter squared, then by multiplying that by the number of square meters in the field (200), the an approximate total population can be calculated (2 million blades of grass).

individuals / sample area = density

density * total area = total population

10,000 * 200 = 2,000,000

Limitations of Using Quadrats

Quadrats give accurate results when the density of individuals does not change much from one location to another in the species' habitat. If one area of the habitat can sustain twice the number of individuals per unit of area than another area in the habitat, then it can drastically bias the results. Avoiding this is relatively simple though. By dividing the habitat into smaller areas with similar populations densities, then figuring out the size of the populations in those areas, and combining them, a more accurate model of the population of the total area can be calculated.

The Capture-mark-release-recapture Method of Calculating Populations

Calculating the population size of motile creatures can be extremely difficult. As they are able to move, one cannot section off their habitat into a number of quadrats, and then count the number in each quadrat, as the animals being counted may simply move out of the quadrat. Furthermore, it can be very difficult to accurately gauge how many individuals are being counted. One would have to use some sort of distinct marking to make sure not to count certain individuals twice.

The Method to this Madness

A very clever method of counting motile populations has been devised. It is known as the Capture-Mark-Release-Recapture method. The procedure is very simple: capture a number of individuals of species who's population is being calculated, mark them so that each they can be later identified as separate from those that have not been captured, then release them. After waiting enough time to allow the marked sample to disperse back into its habitat, another sample of the same species is captured. By calculating the percentage of marked individuals in the second capture group, and knowing the total number of individuals that were marked, it is relatively simple to approximate the total number of individuals in a population.

For Example:

To figure out the number of elephants in India, a research team captures, marks, and releases 500 elephants. Several weeks later, after allowing the marked elephants to recirculate into the unmarked population, another 500 are captured. If 50 of the newly captured elephants are marked (10% of the second sample), this suggests that around 10% of the total elephants in India were previously captured and marked. Since the first sample was of 500 elephants, and that is only 10% of the population, it is reasonable to expect that the total population of the elephants in India is close to 5000.

total marked / total individuals = sample marked / sample total

500 / total = 50 / 500

500 * 500 / 50 = total

total = 5000

Drawbacks & Limitations

The Mark-Recapture method (as it is often abbreviated) of calculating population sizes, although powerful, is only accurate under certain assumptions - that the population size of both marked and unmarked individuals stays the same, and that the second sample accurately portrays the ratio of marked to unmarked individuals. This means that to stay accurate there must be no immigration, emigration, deaths, births or shed markings, and that marking does not increase the ease of further captures. While many factors cannot be isolated, two things can be done to best avoid bias: waiting only enough time between captures so that the marked individuals can evenly disperse, and properly designing the marking method.

Avoiding Bias

Properly designed markings are crucial. They must be secure enough to not fall off or wear out. The markings must be obvious enough so that they are not missed or confused with other possible markings while counting the recapture sample, but must be discreet enough so that the marked individual is allowed to re-enter its natural habitat. The marking process and the mark itself must not change the likelihood of recapturing marked individuals (for example, some mouse populations used to be captured with bait. Some mice would become "trap shy" once captured, understanding that the food was a bait, others would become "trap friendly" once captured, returning multiple times to secure a free meal.) Lastly, the markings must not damage the creature's ability to survive. Although these requirements may seem difficult to fulfill, properly designed markings can greatly decrease the possibility of error in the counting.

Properly timing the recapture is also necessary. Enough time must be allowed so that the marked sample can disperse into the unmarked population, but, with more time, the probability of births, deaths, emigration and immigration increases. Calculating the best time is a tricky business of minimizing the amount of population change, while maximizing the amount of time allowing marked individuals to disperse.

Limits to Population Growth

No population can increase without limitation. Instead, populations in natural ecosystems increase or decrease in response to the changes in the factors that restrict growth. Many factors influence population densities and growth, and these factors may lead to oscillations in population size over time. It is also often difficult to determine the exact factor limiting growth. Many different factors may combine to produce unexpected results.

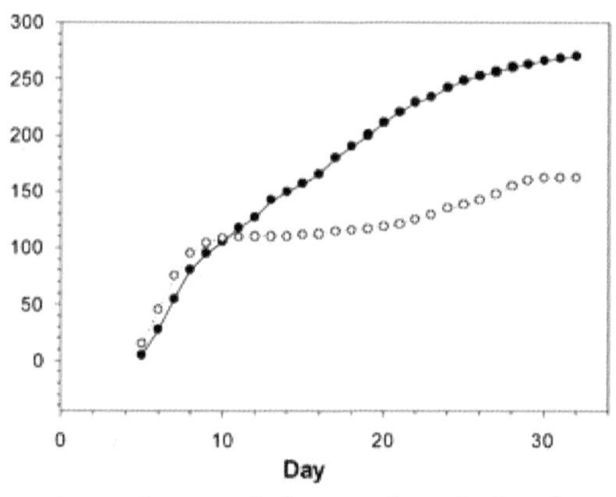

Figure: Lemming population numbers during the period

One organism that experiences rapid oscillations in population density in response to growth limiting factors is the lemming. Lemmings are small rodents that live in the high-Arctic tundra of Greenland and in other arctic environments across the world. One species, the collard lemming Dycrostonyx groenlandicus, is a chubby-looking rodent living in the arctic of North America and Greenland. It is food for a number of vertebrate predators, including the stoat a short-tailed weasel, the arctic fox, the snowy owl, and the long-tailed skua a seabird). Because of the simplicity of this system, lemming population dynamics make an excellent case study for examining the factors regulating population growth. Gilg et al. 2003 studied this system in the Karup Valley of northeast Greenland. The lemming population increased and decreased in a regular four-year cycle during the study period, 1988–2002. The number of lemmings increased to as many as ten per hectare. Gilg et al. 2003 found that the single most important factor limiting lemming population size was the predation pressure affecting those populations. The owl, fox, and skua switched to lemming predation as the lemming numbers increased, preventing rapid population growth. As the lemmings provided the stoat with additional food, their reproductive success increased, allowing an increased stoat population. Stoat population expansion eventually overran lemming population growth, and the lemming population collapsed, soon followed by a collapse in the stoat population, and the cycle repeated itself.

Density Dependant Limitation

Limitations to population growth are either density-dependant or density-independent. Density-dependent factors include disease, competition, and predation. Density-dependant factors can have either a positive or a negative correlation to population size. With a positive relationship, these limiting factors increase with the size of the population and limit growth as population size increases. With a negative relationship, population growth is limited at low densities and becomes less limited as it grows.

Density-dependant factors may influence the size of the population by changes in reproduction or survival. Wauters & Lens 1995 studied how food availability and density combine to limit red squirrel reproduction rates. The red squirrel Sciurus vulgaris is a small rodent inhabiting forests in Europe and Asia. They studied squirrels in both coniferous and deciduous woodlands and investigated how limitations in food resulted in limitations in reproduction as population densities increased. They found that when squirrel densities were high, territoriality relegated some females to poor quality territory, which in turn reduced their reproductive success. When squirrel densities were low, no females occupied the low-quality territory. Thus, it was not all individuals suffering from reduced ability to reproduce e.g., fecundity due to the density increase. Instead, a greater proportion of the population was living in poor-quality habitat, while those still living in good habitat continued to have success. This in turn led to a decrease in per capita birth rate, a limitation in population growth as a function of population density.

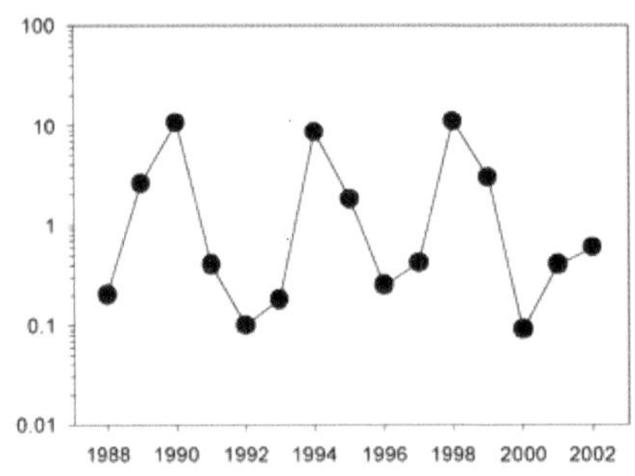

Figure: Mortality of juvenile red deer from zero to 24 months of age (hollow circles) and mortality of adults between 5–10 years old.

Density dependant factors may also affect population mortality and migration. Clutton-Brock et al. found these density-dependant controls in a population of red deer Cervus elaphus in the Scottish Highlands. Both juvenile and adult mortality was significantly affected by population density, with juvenile mortality more strongly influenced than adult mortality. Furthermore, they found that these differences were stronger among males than females, so that increasing population density caused a shift in the sex ratio of females to males. This effect was enhanced by decreased male immigration and increased male emigration. Thus, density-dependant controls on population growth not only increased with increasing density, but also differentially affected males and females within the population.

Density Independent Limitation

Factors that decrease population growth can be defined as environmental stress including limitations in food, predation, and other density-dependant factors Sibley & Hone 2002. However, many sources of environmental stress affect population growth, irrespective of the density of the population. Density-independent factors, such as environmental stressors and catastrophe, are not influenced by population density change. While the previously mentioned density-dependant factors are often biotic, density-independent factors are often abiotic. These density-independent factors include food or nutrient limitation, pollutants in the environment, and climate extremes, including seasonal cycles such as monsoons. In addition, catastrophic factors can also impact population growth, such as fires and hurricanes.

The quality of nutrients e.g., food quality, amount of particular plant nutrients) in an environment affects the ability of an organism to survive, grow, and reproduce. The lower the quality of the nutrients, the higher the environmental stress. In the freshwater Laurentian Great Lakes, particularly in Lake Erie, the factor limiting algal growth was found to be phosphorus. David Schindler and his colleagues at the Experimental Lakes Area Ontario, Canada demonstrated that phosphorus was the growth-limiting factor in temperate North American lakes using whole-lake treatment and controls. This work encouraged the passage of the Great Lakes Water Quality Agreement of 1972 — a reduction in phosphorus load from municipal sources was predicted to lead to a corresponding reduction in the total algal biomass and harmful cyanobacterial (blue-green algae) blooms (figure). As annual phosphorus loads decreased in the mid 1980s, there was some indication that Lake Erie was improving in terms of decreased total phytoplankton (photosynthetic algae and cyanobacteria) biomass Makarewicz 1993. Further improvement continued until the mid 1990s, until an introduced species, the zebra mussel, began altering the internal phosphorus dynamics of the lake by mineralization excretion of digested algae (figure below).

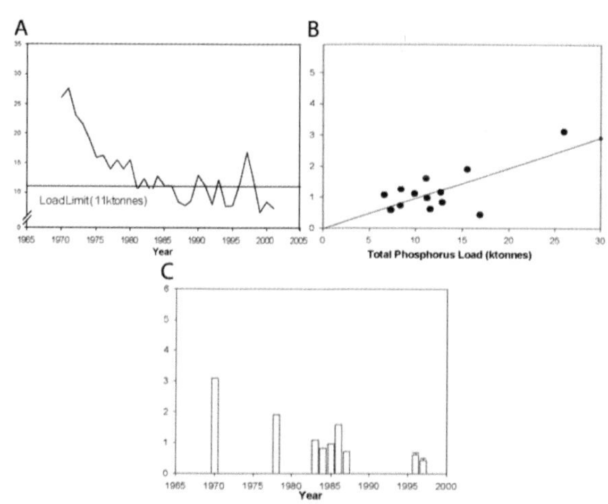

Figure: Temporal dynamics of lake-wide annual total phosphorus load.

A Phosphorus load decreased toward the target annual load 11 ktonnes. B Lake Erie seasonal average phytoplankton biomass (mg/L^{-1}) as a function of lake-wide annual estimated total phosphorus loading ktonnes for the central basin (slope = 0.0972, r2 = 0.49, p = 0.007). C Change in Lake Erie seasonal average phytoplankton biomass in the central.

Pollutants also contribute to environmental stress, limiting the growth rates of populations. Although each species has specific tolerances for environmental toxins, amphibians in general are particularly susceptible to pollutants in the environment. For example, pesticides and other endocrine disrupting toxins can strongly control the growth of amphibians. These chemicals are used to control agricultural pests but also run into freshwater streams and ponds where amphibians live and breed. They affect the amphibians both with direct increases in mortality and indirect limitation in growth, development, and reduction in fecundity. Rohr, found, among many other examples, that these compounds affect salamander embryo survival in affected ponds, increased deformities, and delayed development and growth, lengthening their vulnerability to predators by remaining small sized for longer periods. These effects limit population growth irrespective of the size of the amphibian population and are not limited to pesticides but also include pH and thermal pollution, herbicides, fungicides, heavy metal contaminations, etc.

Environmental catastrophes such as fires, earthquakes, volcanoes and floods can strongly affect population growth rates via direct mortality and habitat destruction. A large-scale natural catastrophe occurred in 2005 when hurricane Katrina impacted the coastal regions of the Gulf of Mexico in the southern United States. Katrina altered habitat for coastal vegetation by depositing more than 5 cm of sediment over the entire coastal wetland zone. In these areas, substantial improvement in the quality of wetlands for plant growth occurred after many years of wetland loss due to control of the Mississippi River flow. At the same time, however, almost 100 km^2 of wetland was destroyed and converted to open sea, completely eliminating wetland vegetation. More recently the Gulf oil spill in 2010 has again impacted the coastal wetland vegetation. Though human derived, this large-scale environmental disaster will have long-term impacts on the population growth of not only vegetation but all organisms in the wetlands and nearshore regions of the Gulf of Mexico.

Exponential Growth

Populations with unlimited natural resources grow very rapidly, and then population growth decreases as resources become depleted. This accelerating pattern of increasing population size is called exponential growth.

The best example of exponential growth is seen in bacteria. Bacteria reproduce by prokaryotic fission. This division takes about an hour for many bacterial species. If 1000 bacteria are placed in a large flask with an unlimited supply of nutrients (so the nutrients will not become depleted), after an hour, there is one round of division and each organism divides, resulting in 2000 organisms—an increase of 1000. In another hour, each of the 2000 organisms will double, producing 4000, an increase of 2000 organisms. After the third hour, there should be 8000 bacteria in the flask, an increase of 4000 organisms. The important concept of exponential growth is the accelerating population growth rate—the number of organisms added in each reproductive generation—that is, it is increasing at a greater and greater rate. After 1 day and 24 of these cycles, the population would have increased from 1000 to more than 16 billion. When the population size, N, is plotted over time, a J-shaped growth curve is produced (figure).

The bacteria example is not representative of the real world where resources are limited. Furthermore, some bacteria will die during the experiment and thus not reproduce, lowering the growth rate. Therefore, when calculating the growth rate of a population, the death rate (D) (number organisms that die during a particular time interval) is subtracted from the birth rate (B) (number organisms that are born during that interval). This is shown in the following formula:

$$\frac{\Delta N\,(change\,in\,number)}{\Delta T\,(change\,in\,time)} = B(birth\,rate) - D(death\,rate)$$

The birth rate is usually expressed on a per capita (for each individual) basis. Thus, B (birth rate) = bN (the per capita birth rate "b" multiplied by the number of individuals "N") and D (death rate) = dN (the per capita death rate "d" multiplied by the number of individuals "N"). Additionally, ecologists are interested in the population at a particular point in time, an infinitely small time interval. For this reason, the terminology of differential calculus is used to obtain the "instantaneous" growth rate, replacing the change in number and time with an instant-specific measurement of number and time.

$$\frac{dN}{dT} = bN - dN = (b-d)N$$

Notice that the "d" associated with the first term refers to the derivative (as the term is used in calculus) and is different from the death rate, also called "d." The difference between birth and death rates is further simplified by substituting the term "r" (intrinsic rate of increase) for the relationship between birth and death rates:

$$\frac{dN}{dT} = rN$$

The value "r" can be positive, meaning the population is increasing in size; or negative, meaning the population is decreasing in size; or zero, where the population's size is unchanging, a condition known as zero population growth. A further refinement of the formula recognizes that different species have inherent differences in their intrinsic rate of increase (often thought of as the potential for reproduction), even under ideal conditions. Obviously, a bacterium can reproduce more rapidly and have a higher intrinsic rate of growth than a human. The maximal growth rate for a species is its biotic potential, or r_{max}, thus changing the equation to:

$$\frac{dN}{dT} = r_{max}N$$

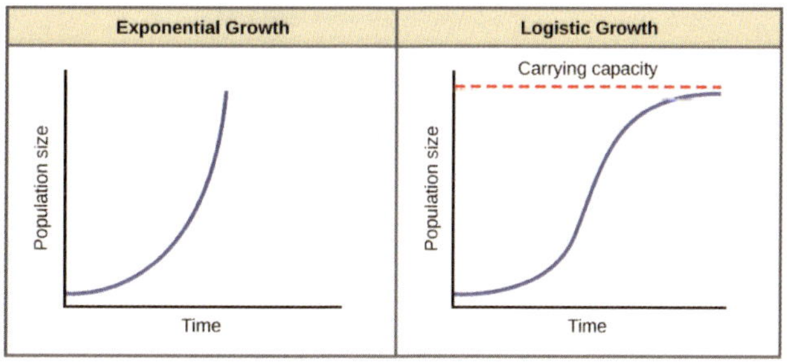

When resources are unlimited, populations exhibit exponential growth, resulting in a J-shaped curve. When resources are limited, populations exhibit logistic growth. In logistic growth, population expansion decreases as resources become scarce, and it levels off when the carrying capacity of the environment is reached, resulting in an S-shaped curve.

Logistic Growth

Exponential growth is possible only when infinite natural resources are available; this is not the case in the real world. Charles Darwin recognized this fact in his description of the "struggle for existence," which states that individuals will compete (with members of their own or other species) for limited resources. The successful ones will survive to pass on their own characteristics and traits (which we know now are transferred by genes) to the next generation at a greater rate: a process known as natural selection. To model the reality of limited resources, population ecologists developed the logistic growth model.

Carrying Capacity and the logistic Model

In the real world, with its limited resources, exponential growth cannot continue indefinitely. Exponential growth may occur in environments where there are few individuals and plentiful resources, but when the number of individuals becomes large enough, resources will be depleted, slowing the growth rate. Eventually, the growth rate will plateau or level off. This population size, which represents the maximum population size that a particular environment can support, is called the carrying capacity, or K.

The formula we use to calculate logistic growth adds the carrying capacity as a moderating force in the growth rate. The expression "K – N" is indicative of how many individuals may be added to a population at a given stage, and "K – N" divided by "K" is the fraction of the carrying capacity available for further growth. Thus, the exponential growth model is restricted by this factor to generate the logistic growth equation:

$$dN/dT = r\max * (dN/dT) = r\max * N * ((KN)/K)$$

Notice that when N is very small, (K-N)/K becomes close to K/K or 1; the right side of the equation reduces to $r_{max}N$, which means the population is growing exponentially and is not influenced by carrying capacity. On the other hand, when N is large, (K-N)/K come close to zero, which means that population growth will be slowed greatly or even stopped. Thus, population growth is greatly slowed in large populations by the carrying capacity K. This model also allows for negative population growth or a population decline. This occurs when the number of individuals in the population exceeds the carrying capacity (because the value of (K-N)/K is negative).

A graph of this equation yields an S-shaped curve; it is a more-realistic model of population growth than exponential growth. There are three different sections to an S-shaped curve. Initially, growth is exponential because there are few individuals and ample resources available. Then, as resources begin to become limited, the growth rate decreases. Finally, growth levels off at the carrying capacity of the environment, with little change in population size over time.

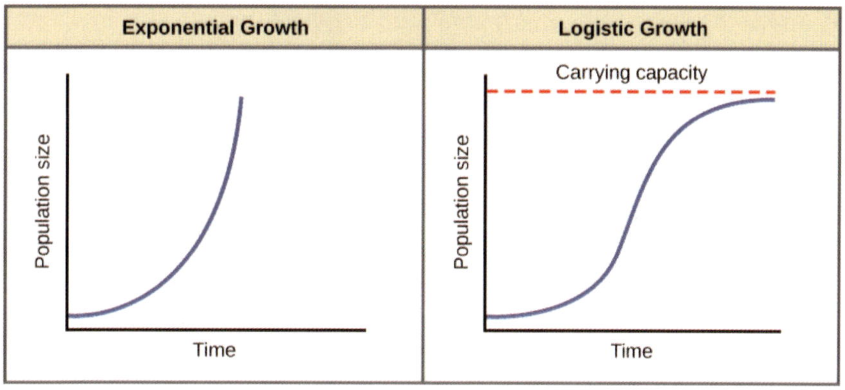

Exponential and logistical population growth: When resources are unlimited, populations exhibit exponential growth, resulting in a J-shaped curve. When resources are limited, populations exhibit logistic growth. In logistic growth, population expansion decreases as resources become scarce, leveling off when the carrying capacity of the environment is reached, resulting in an S-shaped curve.

Role of Intraspecific Competition

The logistic model assumes that every individual within a population will have equal access to resources and, thus, an equal chance for survival. For plants, the amount of water, sunlight, nutrients, and the space to grow are the important resources, whereas in animals, important resources include food, water, shelter, nesting space, and mates.

In the real world, the variation of phenotypes among individuals within a population means that some individuals will be better adapted to their environment than others. The resulting competition between population members of the same species for resources is termed intraspecific competition (intra- = "within"; -specific = "species"). Intraspecific competition for resources may not affect populations that are well below their carrying capacity as resources are plentiful and all individuals can obtain what they need. However, as population size increases, this competition intensifies. In addition, the accumulation of waste products can reduce an environment's carrying capacity.

Examples of Logistic Growth

Yeast, a microscopic fungus used to make bread and alcoholic beverages, exhibits the classical S-shaped curve when grown in a test tube (a). Its growth levels off as the population depletes the nutrients that are necessary for its growth. In the real world, however, there are variations to this idealized curve. Examples in wild populations include sheep and harbor seals (b). In both examples, the population size exceeds the carrying capacity for short periods of time and then falls below the carrying capacity afterwards. This fluctuation in population size continues to occur as the population oscillates around its carrying capacity. Still, even with this oscillation, the logistic model is confirmed.

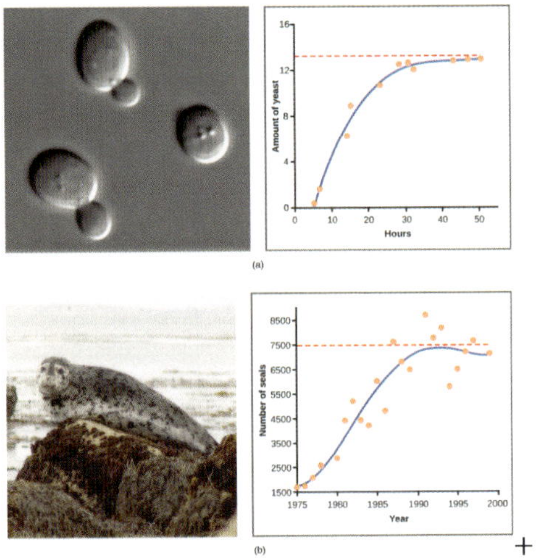

Logistic population growth: (a) Yeast grown in ideal conditions in a test tube show a classical S-shaped logistic growth curve, whereas (b) a natural population of seals shows real-world fluctuation.

References

- Population-ecology, science: britannica.com, Retrieved 10 June 2018
- Population-density-definition: statisticshowto.com, Retrieved 26 April 2018
- Calculate-population-density-5371741: sciencing.com, Retrieved 12 March 2018
- Age-structure-definition-3026043: thoughtco.com, Retrieved 16 June 2018
- library, knowledge, Population-limiting-factors-17059572: nature.com, Retrieved 20 March 2018
- Environmental-limits-to-population-growth, boundless-biology: lumenlearning.com, Retrieved 14 May 2018

7

Sub-disciplines of Biology

Biology is a vast subject that branches out into significant sub-disciplines, which makes it essential to understand its various sub-domains. This chapter discusses these branches in extensive detail such as botany, zoology, marine biology, molecular biology, etc. for a holistic understanding of the subject.

Theoretical Biology

Theoretical biology is the study of biology from first principles. The goal for me is to make a full mathematical model of the behavior of a cell which is capable of predicting the behavior in a general situation, at least statistically (meaning, the range of possible behaviors).

The idea is a little difficult to make clear, because there are two immediate things one thinks:

1. Isn't this just physics

Physics will predict the behavior of the atoms in the cell, so it solves the problem of biology in principle. But the biology is not just the physics, because the physics is absolutely enormous, and mostly irrelevant. Which water molecule is where is just not an interesting biological question, and at most contributes only a little stochasticity to the biologically relevant events. Removing information is the main idea.

2. Isn't it intractible

The other respose is that the problem is impossible, because there is no meaningful simplification to be made. The cell is in some sense maximally complex, it is Turing complete. A Turing complete system admits no reduction in description beyond a certain point.

Turing completeness meant that you couldn't reduce the cell to a simple iteration system, like the renormalization group reduces the Ising model. It meant that any simplification of the sort that physicists like, the kind that makes the behavior obvious in a simple model with tractable behavior, was going to be futile in biology, because it would be ignoring the Turing complete aspects.

But then, you can see that we have a whole field that is devoted to artificial Turing complete systems, and they make progress. This is computer science. How do they make theory. To a large extent, their theory consists of ways of making a language which more succinctly expresses the structure of algorithms and their properties.

So proceeding in this way, you can make a language for the information transformations in the cell, and so long as you keep making the language higher level in the proper way, you will produce a more and more elegant description of the program of the cell. This is what theoretical biology means: it is the program of describing the computational algorithms the cell is performing with a more and more high level Turing complete language appropriate to the domain. When you do it right, you have a minimal description, which captures the function in a minimal number of bits. At this point, you can start to make predictions about behavior, and produce artificial biological circuits comparable to natural ones.

The major lesson from this is that proteins are computationally peripheral outside of bacterial cells, and the major story computation wise is in noncoding purely computing RNA.

Models in Theoretical Biology

Theoretical approaches are gaining in prominence throughout the biological sciences, contributing to what can, without exaggeration, be called a renaissance of the oretical biology. Models and modelling strategies are an important part of this trend. Indeed, in many instances, developments in theory are a direct outgrowth of modelling efforts. However, the relationship between models and theories, and be-tween modelling and theoretical biology, are complex and manifold. Following Lau-bichler, we distinguish at least four different, yet interrelated, areas of theoretical biology in the post genomic age. Different types of modelling and modelling strategies define the distinctions between these areas.

The first dimension of theoretical biology reflects the need to analyze, organize ,and manage large amounts of data. This is the domain of bioinformatics and of certain areas of computational biology, which deal with the substantial challenges of data analysis and representation. The problem here is not only that biology is currently experiencing a phase of exponential growth in the amount of data available; there are also different kinds of data that need to be connected (integrating sequence information and medical histories, for instance). Consequently, data mining is now a-days a desired skill for theoretical and experimental biologists alike. But many of the bioinformatics tools, which the average user simply wants to apply to analyze data, actually represent heuristic and theoretical models of the underlying complexities of biological systems.

Even though one can look at the computational and mathematical tools developed in bioinformatics as the microscope of the twenty-first century, allowing us to "see" new connections and structures, these nevertheless embody more conceptual and theory-driven dimensions of biological research and are in fact models that represent certain aspects of biological systems. A DNA chip and the computational tools that enable us to analyze molecular data are, in a way, a model of the transcriptional state of a cell or a tissue. These and other aspects of merging new data-rich fields with traditional modelling strategies are particularly relevant in the emerging field of systems biology.

A second area of theoretical biology—for many, the primary aspect of the oreticalbiology—involves mathematical and computational model-building and analysis. Over the last decades, simulation and modelling approaches have increased dramatically. Today these routinely include standard systems of differential equations, agent-based models, and spatial models. All this has been made possible by the substantial increase in computational capacity. While model-building and simulation have become an integral part of most areas of biology, they also remain a core part of theoretical biology proper.

In our view, the main emphasis of theoretical biology is not so much on creating an accurate mathematical representation of a particular event, but rather on gaining insights of a more general nature by means of abstract representations of biological processes. These representations are, of course, a form of model of the phenomenon in question. Only from the perspective of such a model is it possible to inquire towhat extent different phenomena can be understood as instances of similar underlying processes or dynamics. This is a primary goal of theoretical biology.

A focus on concept formation and conceptual analysis constitutes the third area of theoretical biology. While this is an important and sometimes overlooked part of bi-ology, it is also related to many ongoing efforts within the philosophy of biology. Conceptual analysis is often closely linked to mathematical analysis, including several mathematical modelling strategies—even more so as many theoretical ideas in biology become expressed in formal terms. However, conceptual analysis is by no means restricted to mathematical models. As several examples in this volume demonstrate, it is precisely through the application of a concept, such as constraint or optimization, to a wide range of phenomena that new insights will be gained.

Darwin's own work, for instance, contains several examples of this kind. Probably the best-known one is the conceptual model provided by artificial selection, which he knew from first hand experience and then employed it to infer processes in natural populations. We submit—and the chapters in this volume.

The fourth area of theoretical biology involves theory integration. This function arises from the fact that formerly distinctly separate disciplines and research domains have begun to merge, both methodologically and conceptually. Theory integration requires different types of models, most prominently heuristic and conceptual models. The goal is to arrive at formal and theoretical models that capture the essential features of a wide range of biological phenomena. Theory integration is often connected to the idea of a "synthesis." One prominent example of this kind of integration is evolutionary developmental biology (EvoDevo), which combines several fields of investigation in a practical and theoretical sense, leading to new research strategies and agendas. As the discussion in chapter 16 of this volume shows, achieving such a synthesis depends in crucial ways on the avail-ability of the right kinds of models, both conceptual and material. It is all the more important today because the life sciences are in a phase of rapid expansion and ever increasing specialization, while at the same time biological theories and concepts have become increasingly more widespread outside the traditional boundaries of the life sciences (from evolutionary psychology to artificial life).

Areas of Research

Several areas of specialized research in mathematical and theoretical biology as well as external links to related projects in various universities are concisely presented in the following subsections, including also a large number of appropriate validating references from a list of several thousands of published authors contributing to this field. Many of the included examples are characterised by highly complex, nonlinear, and supercomplex mechanisms, as it is being increasingly recognised that the result of such interactions may only be understood through a combination of mathematical, logical, physical/chemical, molecular and computational models.

Evolutionary Biology

Ecology and evolutionary biology have traditionally been the dominant fields of mathematical biology.

Evolutionary biology has been the subject of extensive mathematical theorizing. The traditional approach in this area, which includes complications from genetics, is population genetics. Most population geneticists consider the appearance of new alleles by mutation, the appearance of new genotypes by recombination, and changes in the frequencies of existing alleles and genotypes at a small number of gene loci. When infinitesimal effects at a large number of gene loci are considered, together with the assumption of linkage equilibrium or quasi-linkage equilibrium, one derives quantitative genetics. Ronald Fisher made fundamental advances in statistics, such as analysis of variance, via his work on quantitative genetics. Another important branch of population genetics that led to the extensive development of coalescent theory is phylogenetics. Phylogenetics is an area that deals with the reconstruction and analysis of phylogenetic (evolutionary) trees and networks based on inherited characteristics Traditional population genetic models deal with alleles and genotypes, and are frequently stochastic.

Many population genetics models assume that population sizes are constant. Variable population sizes, often in the absence of genetic variation, are treated by the field of population dynamics. Work in this area dates back to the 19th century, and even as far as 1798 when Thomas Malthus formulated the first principle of population dynamics, which later became known as the Malthusian growth model. The Lotka–Volterra predator-prey equations are another famous example. Population dynamics overlap with another active area of research in mathematical biology: mathematical epidemiology, the study of infectious disease affecting populations. Various models of the spread of infections have been proposed and analyzed, and provide important results that may be applied to health policy decisions.

In evolutionary game theory, developed first by John Maynard Smith and George R. Price, selection acts directly on inherited phenotypes, without genetic complications. This approach has been mathematically refined to produce the field of adaptive dynamics.

Computer Models and Automata Theory

A monograph on this topic summarizes an extensive amount of published research in this area up to 1986, including subsections in the following areas: computer modeling in biology and medicine, arterial system models, neuron models, biochemical and oscillation networks, quantum automata, quantum computers in molecular biology and genetics, cancer modelling, neural nets, genetic networks, abstract categories in relational biology, metabolic-replication systems, category theoryapplications in biology and medicine, automata theory, cellular automata, tessellation models and complete self-reproduction, chaotic systems in organisms, relational biology and organismic theories.

Modeling Cell and Molecular Biology

This area has received a boost due to the growing importance of molecular biology.

- Mechanics of biological tissues.

- Theoretical enzymology and enzyme kinetics.
- Cancer modelling and simulation.
- Modelling the movement of interacting cell populations.
- Mathematical modelling of scar tissue formation.
- Mathematical modelling of intracellular dynamics.
- Mathematical modelling of the cell cycle.

Modelling Physiological Systems

- Modelling of arterial disease.
- Multi-scale modelling of the heart.
- Modelling electrical properties of muscle interactions, as in bidomain and monodomain models.

Molecular Set Theory

Molecular set theory (MST) is a mathematical formulation of the wide-sense chemical kinetics of biomolecular reactions in terms of sets of molecules and their chemical transformations represented by set-theoretical mappings between molecular sets. It was introduced by Anthony Bartholomay, and its applications were developed in mathematical biology and especially in mathematical medicine. In a more general sense, MST is the theory of molecular categories defined as categories of molecular sets and their chemical transformations represented as set-theoretical mappings of molecular sets. The theory has also contributed to biostatistics and the formulation of clinical biochemistry problems in mathematical formulations of pathological, biochemical changes of interest to physiology, clinical biochemistry and medicine.

Mathematical Methods

A model of a biological system is converted into a system of equations, although the word 'model' is often used synonymously with the system of corresponding equations. The solution of the equations, by either analytical or numerical means, describes how the biological system behaves either over time or at equilibrium. There are many different types of equations and the type of behavior that can occur is dependent on both the model and the equations used. The model often makes assumptions about the system. The equations may also make assumptions about the nature of what may occur.

Mathematical Biophysics

The earlier stages of mathematical biology were dominated by mathematical biophysics, described as the application of mathematics in biophysics, often involving specific physical/mathematical models of biosystems and their components or compartments.

Mathematical Descriptions and their Assumptions

Deterministic Processes

A fixed mapping between an initial state and a final state. Starting from an initial condition and moving forward in time, a deterministic process always generates the same trajectory, and no two trajectories cross in state space.

- Difference equations/Maps – discrete time, continuous state space.
- Ordinary differential equations – continuous time, continuous state space, no spatial derivatives.
- Partial differential equations – continuous time, continuous state space, spatial derivatives.
- Logical deterministic cellular automata – discrete time, discrete state space.

Stochastic Processes

A random mapping between an initial state and a final state, making the state of the system a random variable with a corresponding probability distribution.

- Non-Markovian processes – generalized master equation – continuous time with memory of past events, discrete state space, waiting times of events (or transitions between states) discretely occur.
- Jump Markov process – master equation – continuous time with no memory of past events, discrete state space, waiting times between events discretely occur and are exponentially distributed.
- Continuous Markov process – stochastic differential equations or a Fokker-Planck equation – continuous time, continuous state space, events occur continuously according to a random Wiener process.

Spatial Modelling

One classic work in this area is Alan Turing's paper on morphogenesis entitled *The Chemical Basis of Morphogenesis*, published in 1952 in the Philosophical Transactions of the Royal Society.

- Travelling waves in a wound-healing assay.
- Swarming behaviour.
- A mechanochemical theory of morphogenesis.
- Biological pattern formation.
- Spatial distribution modeling using plot samples.
- Turing patterns

Organizational Biology

Theoretical approaches to biological organization aim to understand the interdependence between the parts of organisms. They emphasize the circularities that these interdependences lead to. Theoretical biologists developed several concepts to formalize this idea.

For example, abstract relational biology (ARB) is concerned with the study of general, relational models of complex biological systems, usually abstracting out specific morphological, or anatomical, structures. Some of the simplest models in ARB are the Metabolic-Replication, or (M,R)--systems introduced by Robert Rosen in 1957-1958 as abstract, relational models of cellular and organismal organization.

Other approaches include the notion of autopoiesis developed by Maturana and Varela, Kauffman's Work-Constraints cycles, and more recently the notion of closure of constraints.

Algebraic Biology

Algebraic biology (also known as symbolic systems biology) applies the algebraic methods of symbolic computation to the study of biological problems, especially in genomics, proteomics, analysis of molecular structures and study of genes.

Computational Neuroscience

Computational neuroscience (also known as theoretical neuroscience or mathematical neuroscience) is the theoretical study of the nervous system.

Model Example: The Cell Cycle

The eukaryotic cell cycle is very complex and is one of the most studied topics, since its misregulation leads to cancers. It is possibly a good example of a mathematical model as it deals with simple calculus but gives valid results. Two research groups have produced several models of the cell cycle simulating several organisms. They have recently produced a generic eukaryotic cell cycle model that can represent a particular eukaryote depending on the values of the parameters, demonstrating that the idiosyncrasies of the individual cell cycles are due to different protein concentrations and affinities, while the underlying mechanisms are conserved.

By means of a system of ordinary differential equations these models show the change in time (dynamical system) of the protein inside a single typical cell; this type of model is called a deterministic process (whereas a model describing a statistical distribution of protein concentrations in a population of cells is called a stochastic process).

To obtain these equations an iterative series of steps must be done: first the several models and observations are combined to form a consensus diagram and the appropriate kinetic laws are chosen to write the differential equations, such as rate kinetics for stoichiometric reactions, Michaelis-Menten kinetics for enzyme substrate reactions and Goldbeter–Koshland kinetics for ultrasensitive transcription factors, afterwards the parameters of the equations (rate constants, enzyme efficiency coefficients and Michaelis constants) must be fitted to match observations; when they cannot be fitted the kinetic equation is revised and when that is not possible the wiring diagram is modified. The parameters are fitted and validated using observations of both wild type and mutants, such as protein half-life and cell size.

To fit the parameters, the differential equations must be studied. This can be done either by simulation or by analysis. In a simulation, given a starting vector (list of the values of the variables), the progression of the system is calculated by solving the equations at each time-frame in small increments.

In analysis, the properties of the equations are used to investigate the behavior of the system depending of the values of the parameters and variables. A system of differential equations can be represented as a vector field, where each vector described the change (in concentration of two or more protein) determining where and how fast the trajectory (simulation) is heading. Vector fields can have several special points: a stable point, called a sink, that attracts in all directions (forcing the concentrations to be at a certain value), an unstable point, either a source or a saddle point, which repels (forcing the concentrations to change away from a certain value), and a limit cycle, a closed trajectory towards which several trajectories spiral towards (making the concentrations oscillate).

A better representation, which handles the large number of variables and parameters, is a bifurcation diagram using bifurcation theory. The presence of these special steady-state points at certain values of a parameter (e.g. mass) is represented by a point and once the parameter passes a certain value, a qualitative change occurs, called a bifurcation, in which the nature of the space changes, with profound consequences for the protein concentrations: the cell cycle has phases (partially corresponding to G1 and G2) in which mass, via a stable point, controls cyclin levels, and phases (S and M phases) in which the concentrations change independently, but once the phase has changed at a bifurcation event (Cell cycle checkpoint), the system cannot go back to the previous levels since at the current mass the vector field is profoundly different and the mass cannot be reversed back through the bifurcation event, making a checkpoint irreversible. In particular the S and M checkpoints are regulated by means of special bifurcations called a Hopf bifurcation and an infinite period bifurcation.

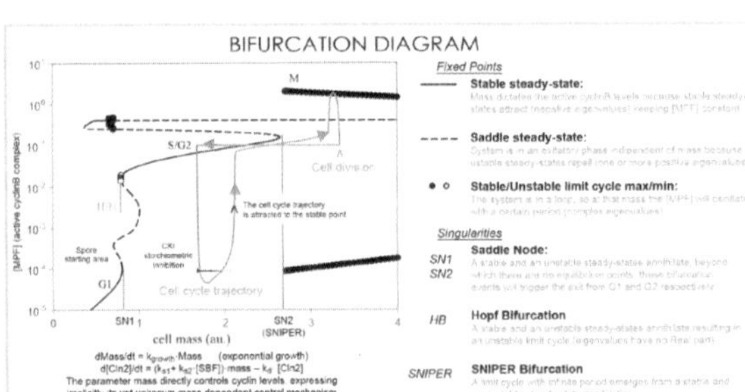

Structural Biology

Structural biology is the study of how biological molecules are built. Using a variety of imaging techniques, scientists view molecules in three dimensions to see how they are assembled, how they function, and how they interact. That has helped researchers understand how the thousands of different molecules in each of our cells work together to keep us healthy. Structural studies have also shown how misshapen molecules make us sick, and as a result, these studies have prompted new treatments for many diseases.

Molecules

Molecules are groups of two or more atoms held together by chemical bonds. Molecules include DNA, RNA, proteins, carbohydrates (sugars), and lipids (fats). Structural biologists are particularly interested in proteins because they do so much of the work in the body. Increasingly, biologists are investigating large molecules made up of combinations of RNA and proteins, called RNA-protein complexes.

Proteins

Proteins are molecules that contribute to virtually every activity in the body. They form hair and fingernails, carry oxygen in the blood, allow muscles to move, and much more.

Top: Protein molecules are made up of amino acids hooked together like beads on a string. Bottom: To become active, proteins must twist and fold into their final, or "native," configuration.

Proteins are made of long strands of small molecules called amino acids. There are 20 amino acids found in nature. Each protein contains a unique combination of a few dozen to many thousands of amino acids. Some proteins consist of multiple amino acid strands wound together.

Even though proteins are strings of amino acids, they do not remain in a straight line. The strands twist, bend, and fold into specific shapes. The way they fold depends in part on the way the amino acids interact with each other. Some sections of proteins form standard "motifs": corkscrew-like coils called alpha helices and flat sections called beta sheets.

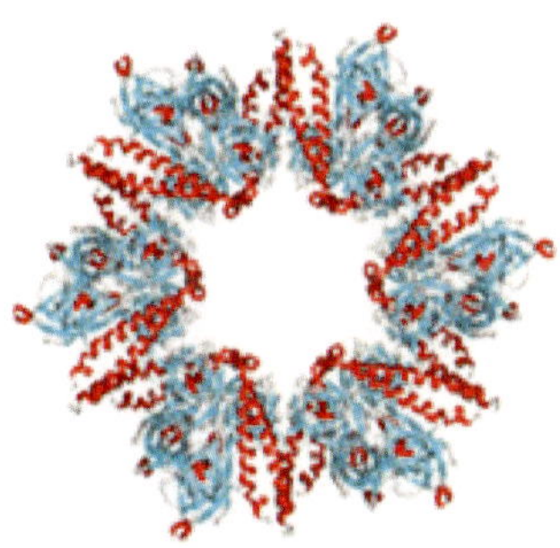

A mix of alpha helices (red; curled ribbon forming a six-sided star) and beta sheets (blue; thinner, tangled strands).

Researchers can easily determine a protein's amino acid sequence. The trick has been figuring out how and why the proteins fold. Scientists are beginning to solve that puzzle with research into how amino acids interact, and they are using powerful new computer programs that help predict protein motifs. Researchers are even starting to design their own brand-new proteins that perform specific jobs. This new work helps scientists understand not only how proteins fold but also how they misfold and malfunction in diseases such as Alzheimer's and cystic fibrosis. Knowing more about these processes might allow researchers to design new treatments.

A protein's structure allows it to perform its job. For instance, antibodies are shaped like a Y. This helps these immune-system proteins bind to foreign molecules such as bacteria or viruses with one end while recruiting other immune-system proteins with the other. DNA polymerase III is donut-shaped. This helps it form a ring around DNA as it copies its genetic information. And proteins called enzymes have grooves and pockets that help them hold onto other molecules to speed chemical reactions. Misfolded, or misshapen, proteins can cause diseases. They often stop working properly and can build up in tissues. Alzheimer's disease, Parkinson's disease, and cystic fibrosis are examples of diseases caused by misfolded proteins.

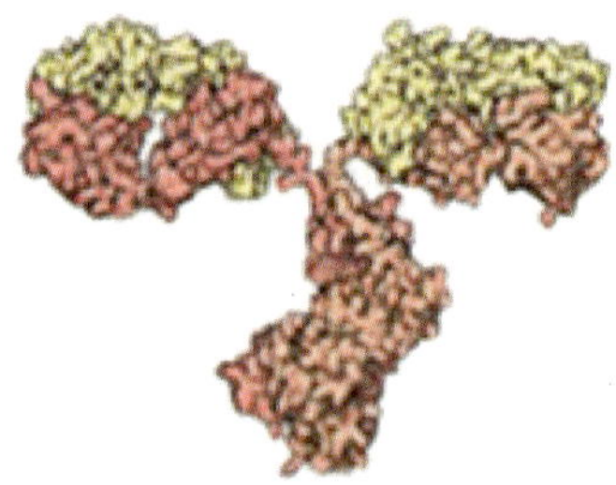

An image of an antibody, an immune system protein that rids the body of foreign material, including bacteria.

There are many different kinds of proteins. For example, many proteins are enzymes that aid biochemical reactions. Others have shapes or specific functions that help cells hold their shape and move, as described in the previous answer. Another type, called transporter proteins, are embedded in the cell's outer membrane and form channels that help vital substances such as sodium or potassium pass into or out of the cell.

Drugs typically work by either blocking or supporting the activity of specific proteins in the body. Using an approach called structure-based drug design, scientists can make a template for a protein and use that blueprint for creating new medicines. They start with a computerized model of the protein structure they're interested in studying. For example, the computer model would allow researchers to examine how two proteins work together. Then, if scientists want to turn off one protein, they would try to design a molecule that would block or alter that interaction.

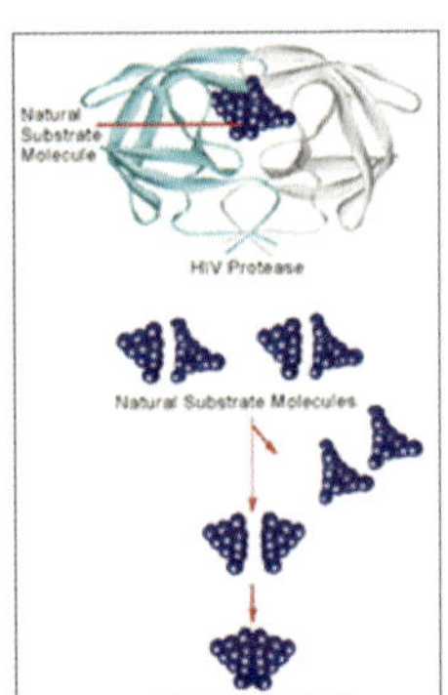

Knowing that HIV protease—an enzyme that breaks down HIV—has two symmetrical halves, pharmaceutical researchers initially attempted to block the enzyme with symmetrical, naturally occurring small molecules. They made these by chopping in half molecules of the natural substrate, then making a new molecule by fusing together two identical halves of the natural substrate.

Example of a Medicine Developed using Structure-based

Researchers used structure-based drug design to develop some anti-HIV drugs. HIV protease is an enzyme that keeps the virus alive. Knowing its structure allowed researchers to determine the kinds of molecules that could stop HIV protease from working. Scientists used computer models to fine tune molecules that could halt virus production. This work led to medicines called protease inhibitors.

Determining Protein Structures

Researchers use several imaging techniques to determine the structure of proteins and other complex molecules. Cryo-electron microscopy (cryo-EM) allows scientists to "see" individual proteins as well as larger structures such as molecular complexes (groups of proteins that combine and function as a unit), viruses, or organelles (specialized structures within the cell that perform specific functions). X-ray crystallography and nuclear magnetic resonance (NMR) spectroscopy also make it possible for researchers to view proteins. To date, researchers have used these techniques to unravel the structure of more than 122,000 proteins.

Cryo-EM

In cryo-EM, researchers rapidly freeze a cell, virus, molecular complex, or other structure so that water molecules do not have time to form crystals. This preserves the sample in its natural state. Scientists use an electron microscope to blast the frozen sample with an electron beam. This creates a two-dimensional projection of the sample on a digital detector. By creating hundreds of projections of the sample from many different angles and then taking the average of these angles, scientists generate a three-dimensional model of its structure. Recent advances in cryo-EM provide highly detailed images of proteins and other biological structures, including larger structures such as RNA-protein complexes.

X-ray Crystallography

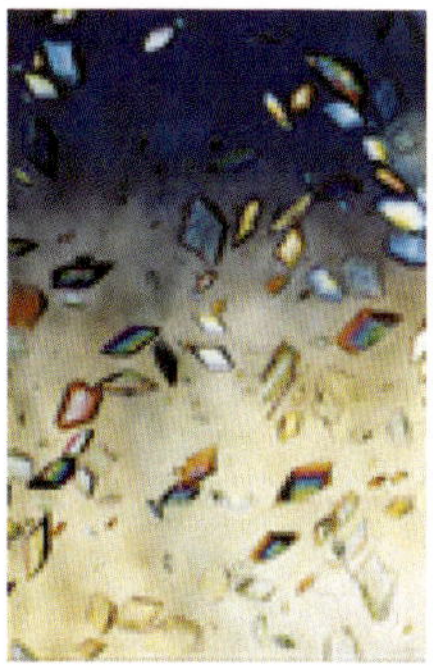

Structural biologists create crystals of proteins, shown here, as step in X-ray crystallography, which can reveal detailed, three-dimensional protein structures.

X-ray crystallography shoots a beam of X-rays through a tiny solid crystal made up of trillions of identical protein molecules. The crystal scatters the X-rays onto an electronic detector, similar to the way images are captured in a digital camera. A computer gauges the intensities of the scattered X-rays to assign a position to each atom in the crystallized molecule. The result is a three-dimensional digital image. This method has been used to determine more than 85 percent of known protein structures.

NMR Spectroscopy

NMR spectroscopy works using the natural magnets—the nuclei of certain atoms—inside proteins. Those natural cellular magnets interact with a big magnet inside the NMR machine. The big magnet forces the protein's magnets to line up. Researchers then blast the sample with a series of split-second radio-wave pulses and observe how the protein's magnets respond. Scientists use several sets of these NMR blasts and combine the data to get a more complete picture of the protein. Although X-ray crystallography can examine larger proteins than NMR is able to do, NMR technology can study proteins immersed in liquid solutions. In contrast, X-ray crystallography requires that proteins be organized into crystals.

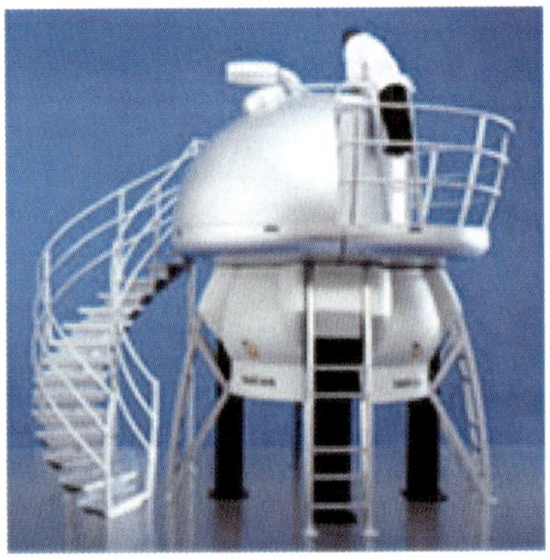

This is a machine used to do NMR spectroscopy. Most of these machines use magnets that are 500 megahertz to 900 megahertz. This magnet is 900 megahertz.

An Insight into Action of Proteins

New technology is beginning to allow researchers to progress from creating static pictures of proteins and other molecules to making movies of their actions.

Images provide snapshots of what these cellular elements are doing at specific points in time. Although they supply valuable information, these still pictures don't capture how proteins and other molecules inside cells are constantly moving and changing, folding and unfolding as they interact. Understanding this dynamic system is critical to unlocking how life works. In addition, there is a whole class of proteins, called intrinsically disordered proteins, that do not hold a specific shape. Their shape adapts to what's going on inside the cell, making it nearly impossible to take a still picture of them.

Scientists are now using powerful computer models to make molecular movies so they can see the full range of proteins in live action. By feeding in information from various imaging techniques about how amino acids and other building blocks interact, scientists can create moving pictures.

Such movies help researchers understand how proteins work in their natural state and allow them to design highly specific drugs.

Future of Structural Biology

Researchers on the frontier of structural biology are merging all of the imaging techniques—X-ray crystallography, NMR, and cryo-EM. This allows them to create a more precise map of what proteins and other molecules look like and how they interact. Scientists can create a single image that zooms in to see specific proteins and also zooms out to see how they interact within the larger cellular structure. In addition to combining existing techniques, scientists are developing ever more powerful methods. For example, new X-ray lasers allow insights into processes that occur in less than one tenth of a trillion of a second, much faster timescales than that captured by other sources of X-rays.

Scientists use super-efficient methods to determine protein structures more quickly than ever before. They also use sophisticated techniques to predict three-dimensional structures of proteins. And they use high-powered computer models to design and create new proteins not found in nature that have useful functions, such as discovering and combating disease. This work will continue to increase our understanding of the diverse roles molecules play in biology and to spur advances in medicine.

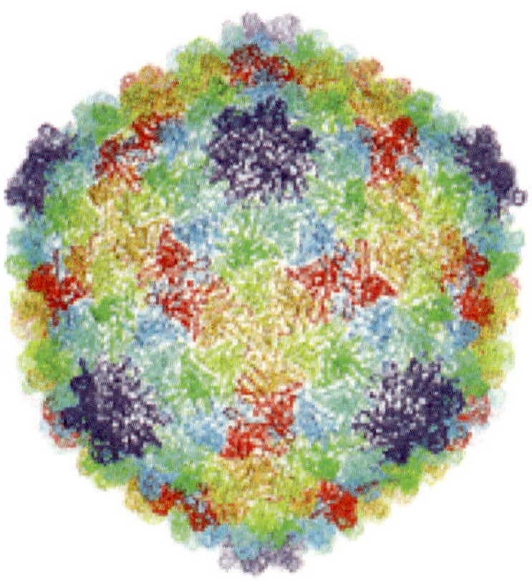

An atomic-scale model of a virus that infects the Salmonella bacterium.

Zoology

Zoology, or animal biology, is the field of biology that involves the study of animals. The word zoology comes from the Greek words zōion, meaning "animal", and logos, meaning "the study of". It encompasses all aspects of scientific knowledge about animals, like embryonic development, evolution, behavior, ecological distribution, and classification. Zoology is broken into many branches because there are so many different ways to study animals; it is also broken into branches based on which animals are being studied.

People have been interested in learning about animals since ancient times. The prominent ancient Greek philosopher Aristotle took detailed notes on animal observations, and inspired other scientists for many hundreds of years. Many universities were founded in Europe in the 16th Century, and by the mid-17th Century, divisions were founded in universities that focused entirely on animal research.

In the 19th Century, the microscope became commonplace in scientific research, and this opened up a whole new realm of possibility; now, the cells of animals could be studied at the microscopic level. Another breakthrough in zoology occurred when the naturalist Charles Darwin developed the theory of evolution by natural selection. This theory revolutionized zoology and taxonomy (classification). More recently, the discovery of DNA as life's genetic material led to even more new research and knowledge about the natural world and the evolutionary relationships between animals.

Areas of Study

Although it is still useful to recognize many disciplines in animal biology—e.g., anatomy or morphology; biochemistry and molecular biology; cell biology; developmental studies (embryology); ecology; ethology; evolution; genetics; physiology; and systematics—the research frontiers occur as often at the interfaces of two or more of these areas as within any given one.

Anatomy or Morphology

Descriptions of external form and internal organization are among the earliest records available regarding the systematic study of animals. Aristotle was an indefatigable collector and dissector of animals. He found differing degrees of structural complexity, which he described with regard to ways of living, habits, and body parts. Although Aristotle had no formal system of classification, it is apparent that he viewed animals as arranged from the simplest to the most complex in an ascending series. Since man was even more complex than animals and, moreover, possessed a rational faculty, he therefore occupied the highest position and a special category. This hierarchical perception of the animate world proved to be useful in every century to the present, except that in the modern view there is no such "scale of nature," and there is change in time by evolution from the simple to the complex.

After the time of Aristotle, Mediterranean science was centred at Alexandria, where the study of anatomy, particularly the central nervous system, flourished and, in fact, first became recognized as a discipline. Galen studied anatomy at Alexandria in the 2nd century and later dissected many animals. Much later, the contributions of the Renaissance anatomist Andreas Vesalius, though made in the context of medicine, as were those of Galen, stimulated to a great extent the rise of comparative anatomy. During the latter part of the 15th century and throughout the 16th century, there was a strong tradition in anatomy; important similarities were observed in the anatomy of different animals, and many illustrated books were published to record these observations.

But anatomy remained a purely descriptive science until the advent of functional considerations in which the correlation between structure and function was consciously investigated; as by French biologists Buffon and Cuvier. Cuvier cogently argued that a trained naturalist could deduce from one suitably chosen part of an animal's body the complete set of adaptations that characterized the organism. Because it was obvious that organisms with similar parts pursue similar habits, they were placed together in a system of classification. Cuvier pursued this viewpoint, which he called the theory of correlations, in a somewhat dogmatic manner and placed himself in opposition to the romantic natural philosophers, such as the German intellectual Johann Wolfgang von Goethe, who saw a tendency to ideal types in animal form. The tension between these schools of thought—adaptation as the consequence of necessary bodily functions and adaptation as an expression of a perfecting principle in nature—runs as a leitmotiv through much of biology, with overtones extending into the early 20th century.

The twin concepts of homology (similarity of origin) and analogy (similarity of appearance), in relation to structure, are the creation of the 19th-century British anatomist Richard Owen. Although they antedate the Darwinian view of evolution, the anatomical data on which they were based became, largely as a result of the work of the German comparative anatomist Carl Gegenbaur, important evidence in favour of evolutionary change, despite Owen's steady unwillingness to accept the view of diversification of life from a common origin.

Anatomy moved from a purely descriptive phase as an adjunct to classificatory studies, into a partnership with studies of function and became, in the 19th century, a major contributor to the concept of evolution.

Taxonomy or Systematics

Not until the work of Carolus Linnaeus did the variety of life receive a widely accepted systematic treatment. Linnaeus strove for a "natural method of arrangement," one that is now recognizable as an intuitive grasp of homologous relationships, reflecting evolutionary descent from a common ancestor; however, the natural method of arrangement sought by Linnaeus was more akin to the tenets of idealized morphology because he wanted to define a "type" form as epitomizing a species.

It was in the nomenclatorial aspect of classification that Linnaeus created a revolutionary advance with the introduction of a Latin binomial system: each species received a Latin name, which was not influenced by local names and which invoked the authority of Latin as a language common to the learned people of that day. The Latin name has two parts. The first word in the Latin name for the common chimpanzee, Pan troglodytes, for example, indicates the larger category, or genus, to which chimpanzees belong; the second word is the name of the species within the genus. In addition to species and genera, Linnaeus also recognized other classificatory groups, or taxa (singular taxon), which are still used; namely, order, class, and kingdom, to which have been added family (between genus and order) and phylum (between class and kingdom). Each of these can be divided further by the appropriate prefix of sub- or super-, as in subfamily or superclass. Linnaeus' great work, the Systema naturae, went through 12 editions during his lifetime; the 13th, and final, edition appeared posthumously. Although his treatment of the diversity of living things has been expanded in detail, revised in terms of taxonomic categories, and corrected in the light of continuing work—for example, Linnaeus treated whales as fish—it still sets the style and method, even to the use of Latin names, for contemporary nomenclatorial work.

Linnaeus sought a natural method of arrangement, but he actually defined types of species on the basis of idealized morphology. The greatest change from Linnaeus' outlook is reflected in the phrase "the new systematics," which was introduced in the 20th century and through which an explicit effort is made to have taxonomic schemes reflect evolutionary history. The basic unit of classification, the species, is also the basic unit of evolution—i.e., a population of actually or potentially interbreeding individuals. Such a population shares, through interbreeding, its genetic resources. In so doing, it creates the gene pool—its total genetic material—that determines the biological resources of the species and on which natural selection continuously acts. This approach has guided work on classifying animals away from somewhat arbitrary categorization of new species to that of recreating evolutionary history (phylogeny) and incorporating it in the system of classification. Modern taxonomists or systematists, therefore, are among the foremost students of evolution.

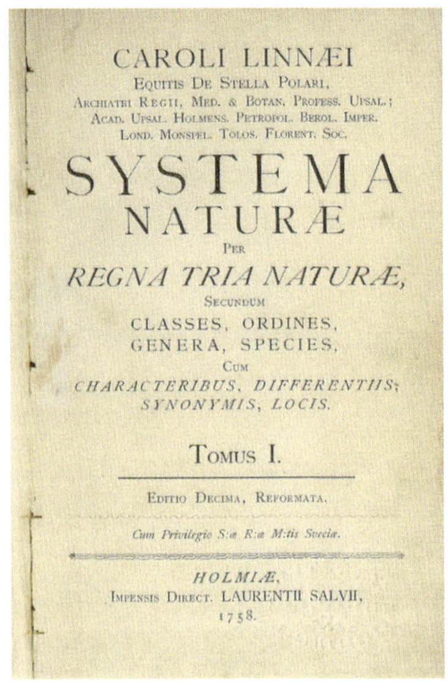
CAROLI LINNÆI
EQUITIS DE STELLA POLARI,
ARCHIATRI REGII, MED. & BOTAN. PROFESS. UPSAL.;
ACAD. UPSAL. HOLMENS. PETROPOL. BEROL. IMPER.
LOND. MONSPEL. TOLOS. FLORENT. SOC.
SYSTEMA
NATURÆ
PER
REGNA TRIA NATURÆ,
SECUNDUM
CLASSES, ORDINES,
GENERA, SPECIES,
CUM
CHARACTERIBUS, DIFFERENTIIS;
SYNONYMIS, LOCIS.

TOMUS I.

EDITIO DECIMA, REFORMATA.

Cum Privilegio S:æ R:æ M:tis Sveciæ.

HOLMIÆ,
IMPENSIS DIRECT. LAURENTII SALVII,
1758.

Linnaeus, Carolus: Systema Naturae

Physiology

The practical consequences of physiology have always been an unavoidable human concern, in both medicine and animal husbandry. Inevitably, from Hippocrates to the present, practical knowledge of human bodily function has accumulated along with that of domestic animals and plants. This knowledge has been expanded, especially since the early 1800s, by experimental work on animals in general, a study known as comparative physiology. The experimental dimension had wide applications following Harvey's demonstration of the circulation of blood. From then on, medical physiology developed rapidly; notable texts appeared, such as Albrecht von Haller's eight-volume work Elementa Physiologiae Corporis Humani (Elements of Human Physiology), which had a medical emphasis. Toward the end of the 18th century the influence of chemistry on physiology became pronounced through Antoine Lavoisier's brilliant analysis of respiration as a form of combustion. This French chemist not only determined that oxygen was consumed by living systems but also opened the way to further inquiry into the energetics of living systems. His studies further strengthened the mechanistic view, which holds that the same natural laws govern both the inanimate and the animate realms.

Physiological principles achieved new levels of sophistication and comprehensiveness with Bernard's concept of constancy of the internal environment, the point being that only under certain constantly maintained conditions is there optimal bodily function. His rational and incisive insights were augmented by concurrent developments in Germany, where Johannes Müller explored the comparative aspects of animal function and anatomy, and Justus von Liebig and Carı Ludwig applied chemical and physical methods, respectively, to the solution of physiological problems. As a result, many useful techniques were advanced—e.g., means for precise measurement of muscular action and changes in blood pressure and means for defining the nature of body fluids.

By this time the organ systems—circulatory, digestive, endocrine, excretory, integumentary, muscular, nervous, reproductive, respiratory, and skeletal—had been defined, both anatomically and functionally, and research efforts were focussed on understanding these systems in cellular and chemical terms, an emphasis that continues to the present and has resulted in specialties in cell physiology and physiological chemistry. General categories of research now deal with the transportation of materials across membranes; the metabolism of cells, including synthesis and breakdown of molecules; and the regulation of these processes.

Interest has also increased in the most complex of physiological systems, the nervous system. Much comparative work has been done by utilizing animals with structures especially amenable to various experimental techniques; for example, the large nerves in squids have been extensively studied in terms of the transmission of nerve impulses, and insect and crustacean eyes have yielded significant information on patterns of sensory inputs. Most of this work is closely associated with studies on animal orientation and behaviour. Although the contemporary physiologist often studies functional problems at the molecular and cellular levels, he is also aware of the need to integrate cellular studies into the many-faceted functions of the total organism.

Embryology, or Developmental Studies

Embryonic growth and differentiation of parts have been major biological problems since ancient times. A 17th-century explanation of development assumed that the adult existed as a miniature—a homunculus—in the microscopic material that initiates the embryo. But in 1759 the German physician Caspar Friedrick Wolff firmly introduced into biology the interpretation that undifferentiated materials gradually become specialized, in an orderly way, into adult structures. Although this epigenetic process is now accepted as characterizing the general nature of development in both plants and animals, many questions remain to be solved. The French physician Marie François Xavier Bichat declared in 1801 that differentiating parts consist of various components called tissues; with the subsequent statement of the cell theory, tissues were resolved into their cellular constituents. The idea of epigenetic change and the identification of structural components made possible a new interpretation of differentiation. It was demonstrated that the egg gives rise to three essential germ layers out of which specialized organs, with their tissues, subsequently emerge. Then, following his own discovery of the mammalian ovum, von Baer in 1828 usefully applied this information when he surveyed the development of various members of the vertebrate groups. At this point, embryology, as it is now recognized, emerged as a distinct subject.

The concept of cellular organization had an effect on embryology that continues to the present day. In the 19th century, cellular mechanisms were considered essentially to be the basis for growth, differentiation, and morphogenesis, or molding of parts. The distribution of the newly formed cells of the rapidly dividing zygote (fertilized egg) was precisely followed to provide detailed accounts not only of the time and mode of germ layer formation but also of the contribution of these layers to the differentiation of tissues and organs. Such descriptive information provided the background for experimental work aimed at elucidating the role of chromosomes and other cellular constituents in differentiation. About 1895, before the formulation of the chromosomal theory of heredity, Theodor Boveri demonstrated that chromosomes show continuity from one cell generation to the next. In fact, biologists soon concluded that in all cells arising from a fertilized egg, half the chromosomes are of maternal and half of paternal origin. The discovery of the constant transmission of the original chromosomal endowment to all cells of the body served to deepen the mystery surrounding the factors that determine cellular differentiation.

The present view is that differential activity of genes is the basis for cellular and tissue differentiation; that is, although the cells of a multicellular body contain the same genetic information, different genes are active in different cells. The result is the formation of various gene products, which regulate the functional and structural differentiation of cells. The actual mechanism involved in the inactivation of certain genes and the activation of others, however, has not yet been established. That cells can move extensively throughout the embryo and selectively adhere to other cells, thus starting tissue aggregations, also contributes to development as does the fate of cells—i.e., certain ones continue to multiply, others stop, and some die.

Research methods in embryology now exploit many experimental situations: both unicellular and multicellular forms; regeneration (replacement of lost parts) and normal development; and growth of tissues outside and inside the host. Hence, the processes of development can be studied with material other than embryos; and the study of embryology has become incorporated into the more inclusive subdiscipline of developmental biology.

Evolutionism

Darwin was not the first to speculate that organisms can change from generation to generation and so evolve, but he was the first to propose a mechanism by which the changes are accumulated. He proposed that heritable variations occur in conjunction with a never-ending competition for survival and that the variations favouring survival are automatically preserved. In time, therefore, the continued accumulation of variations results in the emergence of new forms. Because the variations that are preserved relate to survival, the survivors are highly adapted to their environment. To this process Darwin gave the apt name natural selection.

Many of Darwin's predecessors, notably Jean-Baptiste Lamarck, were willing to accept the idea of species variation, even though to do so meant denying the doctrine of special creation and the static-type species of Linnaeus. But they argued that some idealized perfecting principle, expressed through the habits of an organism, was the basis of variation. The contrast between the romanticism of Lamarck and the objective analysis of Darwin clearly reveals the type of revolution provoked by the concept of natural selection. Although mechanistic explanations had long been available to biologists—forming, for example, part of Harvey's explanation of blood circulation—they did not pervade the total structure of biological thinking until the advent of Darwinism.

There were two immediate consequences of Darwin's viewpoints. One has involved a reappraisal of all subject areas of biology; reinterpretations of morphology and embryology are good examples. The comparative anatomy of the British anatomist Owen became a cornerstone of the evidence for evolution, and German anatomists provided the basis for the comment that evolutionary thinking was born in England but gained its home in Germany. The reinterpretation of morphology carried over into the study of fossil forms, as paleontologists sought and found evidence of gradual change in their study of fossils. But some workers, although accepting evolution in principle, could not easily interpret the changes in terms of natural selection. The German paleontologist Otto Schindewolf, for example, found in shelled mollusks called ammonites evidence of progressive complexity and subsequent simplification of forms. The American paleontologist George Gaylord Simpson, however, has been a consistent interpreter of vertebrate fossils by Darwinian selection. Embryology was seen in an evolutionary light when the German zoologist Ernst Haeckel proposed that the epigenetic sequence of embryonic development (ontogeny) repeated its evolutionary history (phylogeny). Thus, the presence of gill clefts in the mammalian embryo and also in less highly evolved vertebrates can be understood as a remnant of a common ancestor.

The other consequence of Darwinism—to make more explicit the origin and nature of heritable variations and the action of natural selection on them—depended on the emergence of the following: genetics and the elucidation of the rules of Mendelian inheritance; the concept of the gene as the unit of inheritance; and the nature of gene mutation. The development of these ideas provided the basis for the genetics of natural populations.

The subject of population genetics began with the Mendelian laws of inheritance and now takes into account selection, mutation, migration (movement into and out of a given population), breeding patterns, and population size. These factors affect the genetic makeup of a group of organisms that either interbreed or have the potential to do so; i.e., a species. Accurate appraisal of these factors allows precise predictions regarding the content of a given gene pool over significant periods of evolutionary time. From work involving population genetics has come the realization, eloquently documented by two contemporary American evolutionists, Theodosius Dobzhansky and Ernst Mayer, that the species is the basic unit of evolution. The process of speciation occurs as a gene pool breaks up to form isolated gene pools. When selection pressures similar to those of the original gene pool persist in the new gene pools, similar functions and the similar structures on which they depend also persist. When selection pressures differ, however, differences arise. Thus, the process of speciation through natural selection preserves the evolutionary history of a species. The record may be discerned not only in the gross, or macroscopic, anatomy of organisms but also in their cellular structure and molecular organization. Significant work now is carried out, for example, on the homologies of the nucleic acids and proteins of different species.

Genetics

The problem of heredity had been the subject of careful study before its definitive analysis by Mendel. As with Darwin's predecessors, those of Mendel tended to idealize and interpret all inherited traits as being transmitted through the blood or as determined by various "humors" or other vague entities in animal organisms. When studying plants, Mendel was able to free himself of anthropomorphic and holistic explanations. By studying seven carefully defined pairs of characteristics—e.g., tall and short plants; red and white flowers, etc.—as they were transmitted through as many as three successive generations, he was able to establish patterns of inheritance that apply to all sexually reproducing forms. Darwin, who was searching for an explanation of inheritance, apparently never saw Mendel's work, which was simultaneously rediscovered in 1900 by three different European geneticists.

Further progress in genetics was made early in the 20th century, when it was realized that heredity factors are found on chromosomes. The term gene was coined for these factors. Studies by the American geneticist Thomas Hunt Morgan on the fruit fly (Drosophila), moved animal genetics to the forefront of genetic research. The work of Morgan and his students established such major concepts as the linear array of genes on chromosomes; the exchange of parts between chromosomes; and the interaction of genes in determining traits, including sexual differences. In 1927 one of Morgan's former students, Hermann Muller, used X rays to induce the mutations (changes in genes) in the fruit fly, thereby opening the door to major studies on the nature of variation.

Meanwhile, other organisms were being used for genetic studies, most notably fungi and bacteria. The results of this work provided insights into animal genetics just as principles initially obtained from animal genetics provided insight into botanical and microbial forms. Work continues not only on the genetics of humans, domestic animals, and plants but also on the control of development through the orderly regulation of gene action in different cells and tissues.

Cellular and Molecular Biology

Although the cell was recognized as the basic unit of life early in the 19th century, its most exciting period of inquiry has probably occurred since the 1940s. The new techniques developed since that time, notably the perfection of the electron microscope and the tools of biochemistry, have changed the cytological studies of the 19th and early 20th centuries from a largely descriptive inquiry, dependent on the light microscope, into a dynamic, molecularly oriented inquiry into fundamental life processes.

The so-called cell theory, which was enunciated about 1838, was never actually a theory. As Edmund Beecher Wilson, the noted American cytologist, stated in his great work, The Cell,

By force of habit we still continue to speak of the cell 'theory' but it is a theory only in name. In substance it is a comprehensive general statement of fact and as such stands today beside the evolution theory among the foundation stones of modern biology.

More precisely, the cell doctrine was an inductive generalization based on the microscopial examination of certain plant and animal species.

Rudolf Virchow, a German medical officer specializing in cellular pathology, first expressed the fundamental dictum regarding cells in his phrase omnis cellula e cellula (all cells from cells). For cellular reproduction is the ultimate basis of the continuity of life; the cell is not only the basic structural unit of life but also the basic physiological and reproductive unit. All areas of biology were affected by the new perspective afforded by the principle of cellular organization. Especially in conjunction with embryology was the study of the cell most prominent in animal biology. The continuity of cellular generations by reproduction also had implications for genetics. It is little wonder, then, that the full title of Wilson's survey of cytology at the turn of the century was The Cell: Its Role in Development and Heredity.

The study of the cell nucleus, its chromosomes, and their behaviour served as the basis for understanding the regular distribution of genetic material during both sexual and asexual reproduction. This orderly behaviour of the nucleus made it appear to dominate the life of the cell, for by contrast the components of the rest of the cell appeared to be randomly distributed.

The biochemical study of life had helped in the characterization of the major molecules of living systems—proteins, nucleic acids, fats, and carbohydrates—and in the understanding of metabolic processes. That nucleic acids are a distinctive feature of the nucleus was recognized after their discovery by the Swiss biochemist Johann Friedrich Miescher in 1869. In 1944 a group of American bacteriologists, led by Oswald T. Avery, published work on the causative agent of pneumonia in mice (a bacterium) that culminated in the demonstration that deoxyribonucleic acid (DNA) is the chemical basis of heredity. Discrete segments of DNA correspond to genes, or Mendel's hereditary factors. Proteins were discovered to be especially important for their role in determining cell structure and in controlling chemical reactions.

The advent of techniques for isolating and characterizing proteins and nucleic acids now allows a molecular approach to essentially all biological problems—from the appearance of new gene products in normal development or under pathological conditions to a monitoring of changes in and between nerve cells during the transmission of nerve impulses.

Ecology

The harmony that Linnaeus found in nature, which redounded to the glory and wisdom of a Judaeo-Christian god, was the 18th-century counterpart of the balanced interaction now studied by ecologists. Linnaeus recognized that plants are adapted to the regions in which they grow, that insects play a role in flower pollination, and that certain birds prey on insects and are in turn eaten by other birds. This realization implies, in contemporary terms, the flow of matter and energy in a definable direction through any natural assemblage of plants, animals, and microorganisms. Such an assemblage, termed an ecosystem, starts with the plants, which are designated as producers because they maintain and reproduce themselves at the expense of energy from sunlight and inorganic materials taken from the nonliving environment around them (earth, air, and water). Animals are called consumers because they ingest plant material or other animals that feed on plants, using the energy stored in this food to sustain themselves. Lastly, the organisms known as decomposers, mostly fungi and bacteria, break down plant and animal material and return it to the environment in a form that can be used again by plants in a constantly renewed cycle.

The term ecology, first formulated by Haeckel in the latter part of the 19th century as "oecology", referred to the dwelling place of organisms in nature. In the 1890s various European and U.S. scientists laid the foundations for modern work through studies of natural ecosystems and the populations of organisms contained within them.

Animal ecology, the study of consumers and their interactions with the environment, is very complex; attempts to study it usually focus on one particular aspect. Some studies, for example, involve the challenge of the environment to individuals with special adaptations (e.g., water conservation in desert animals); others may involve the role of one species in its ecosystem or the ecosystem itself. Food-chain sequences have been determined for various ecosystems, and the efficiency of the transfer of energy and matter within them has been calculated so that their capacity is known; that is, productivity in terms of numbers of organisms or weight of living matter at a specific level in the food chain can be accurately determined.

In spite of advances in understanding animal ecology, this subject area of zoology does not yet have the major unifying theoretical principles found in genetics (gene theory) or evolution (natural selection).

Ethology

The study of animal behaviour (ethology) is largely a 20th-century phenomenon and is exclusively a zoological discipline. Only animals have nervous systems, with their implications for perception, coordination, orientation, learning, and memory. Not until the end of the 19th century did animal behaviour become free from anthropocentric interests and assume an importance in its own right. The British behaviorist C. Lloyd Morgan was probably most influential with his emphasis on parsimonious explanations—i.e., that the explanation "which stands lower in the psychological scale" must be invoked first. This principle is exemplified in the American Herbert Spencer Jennings' pioneering work in 1906 on The Behavior of Lower Organisms.

The study of animal behaviour now includes many diverse topics, ranging from swimming patterns of protozoans to socialization and communication among the great apes. Many disparate hypotheses have been proposed in an attempt to explain the variety of behavioral patterns found in animals. They focus on the mechanisms that stimulate courtship in reproductive behaviour of such diverse groups as spiders, crabs, and domestic fowl; and on whole life histories, starting from the special attachment of newly born ducks and goats to their actual mothers or to surrogate (substitute) mothers. The latter phenomenon, called imprinting, has been intensively studied by the Austrian ethologist Konrad Lorenz. Physiologically oriented behaviour now receives much attention; studies range from work on conditioned reflexes to the orientation of crustaceans and the location and communication of food among bees; such diversity of material is one measure of the somewhat diffuse but exciting current state of these studies.

General Trends

Zoology has become animal biology—that is, the life sciences display a new unity, one that is founded on the common basis of all life, on the gene pool–species organization of organisms, and on the obligatory interacting of the components of ecosystems. Even as regards the specialized features of animals—involving physiology, development, or behaviour—the current emphasis is on elucidating the broad biological principles that identify animals as one aspect of nature. Zoology has thus given up its exclusive emphasis on animals—an emphasis maintained from Aristotle's time well into the 19th century—in favour of a broader view of life. The successes in applying physical and chemical ideas and techniques to life processes have not only unified the life sciences but have also created bridges to other sciences in a way only dimly foreseen by earlier workers. The practical and theoretical consequences of this trend have just begun to be realized.

Methods in Zoology

Because the study of animals may be concentrated on widely different topics, such as ecosystems and their constituent populations, organisms, cells, and chemical reactions, specific techniques are needed for each kind of investigation. The emphasis on the molecular basis of genetics, development, physiology, behaviour, and ecology has placed increasing importance on those techniques involving cells and their many components. Microscopy, therefore, is a necessary technique in zoology, as are certain physicochemical methods for isolating and characterizing molecules. Computer technology also has a special role in the analysis of animal life. These newer techniques are used in addition to the many classical ones—measurement and experimentation at the tissue, organ, organ system, and organismic levels.

Microscopy

In addition to continuous improvements in the techniques of staining cells, so that their components can be seen clearly, the light used in microscopy can now be manipulated to make visible certain structures in living cells that are otherwise undetectable. The ability to observe living cells is an advantage of light microscopes over electron microscopes; the latter require the cells to be in an environment that kills them. The particular advantage of the electron microscope, however, is its great powers of magnification. Theoretically, it can resolve single atoms; in biology, however, magnifications of lesser magnitude are most useful in determining the nature of structures lying between whole cells and their constituent molecules.

Separation and Purification Techniques

The characterization of components of cellular systems is necessary for biochemical studies. The specific molecular composition of cellular organelles, for example, affects their shape and density (mass per unit volume); as a result, cellular components settle at different rates (and thus can be separated) when they are spun in a centrifuge.

Other methods of purification rely on other physical properties. Molecules vary in their affinity for the positive or negative pole of an electrical field. Migration to or away from these poles, therefore, occurs at different rates for different molecules and allows their separation; the process is called electrophoresis. The separation of molecules by liquid solvents exploits the fact that the molecules differ in their solubility, and hence they migrate to various degrees as a solvent flows past them. This process, known as chromatography because of the colour used to identify the position of the migrating materials, yields samples of extraordinarily high purity.

Radioactive Tracers

Radioactive compounds are especially useful in biochemical studies involving metabolic pathways of synthesis and degradation. Radioactive compounds are incorporated into cells in the same way as their nonradioactive counterparts. These compounds provide information on the sites of specific metabolic activities within cells and insights into the fates of these compounds in both organisms and the ecosystem.

Computers

Computers process information using their own general language, which is able to complete calculations as complex and diverse as statistical analyses and determinations of enzymatically controlled reaction rates. Computers with access to extensive data files can select information associated with a specific problem and display it to aid the researcher in formulating possible solutions. They help perform routine examinations such as scanning chromosome preparations in order to identify abnormalities in number or shape. Test organisms can be electronically monitored with computers, so that adjustments can be made during experiments; this procedure improves the quality of the data and allows experimental situations to be fully exploited. Computer simulation is important in analyzing complex problems; as many as 100 variables, for example, are involved in the management of salmon fisheries. Simulation makes possible the development of models that approach the complexities of conditions in nature, a procedure of great value in studying wildlife management and related ecological problems.

Applied Zoology

Animal-related industries produce food (meats and dairy products), hides, furs, wool, organic fertilizers, and miscellaneous chemical byproducts. There has been a dramatic increase in the productivity of animal husbandry since the 1870s, largely as a consequence of selective breeding and improved animal nutrition. The purpose of selective breeding is to develop livestock whose desirable traits have strong heritable components and can therefore be propagated. Heritable components are distinguished from environmental factors by determining the coefficient of heritability, which is defined as the ratio of variance in a gene-controlled character to total variance.

Another aspect of food production is the control of pests. The serious side effects of some chemical pesticides make extremely important the development of effective and safe control mechanisms. Animal food resources include commercial fishing. The development of shellfish resources and fisheries management (e.g., growth of fish in rice paddies in Asia) are important aspects of this industry.

Systems of Classification

As the science of describing, naming, and classifying living and extinct organisms, taxonomy is integral to the discipline of zoology. The study of animals requires that zoologists are clear on the name, description, and classification of their subjects. In order to standardize such matters, the International Code of Zoological Nomenclature (ICZN) was created. The ICZN is a set of rules in zoology to provide the maximum universality and continuity in classifying animals according to taxonomic judgment.

Animals are one of the major groups of organisms, and are classified as the Kingdom Animalia, or Metazoa. Within this kingdom, a major division is between invertebrates and vertebrates. Invertebrates share the common lack of a trait: a vertebral column, or backbone. About 97 percent of all animal species are invertebrates. Vertebrates are animals with a backbone. With invertebrates, there are more than a dozen phyla, including Porifera (sponges), Cnidaria or Coelenterata (jellyfish, corals), Ctenophora (comb jellies), Mollusca (clams, snails, otopuses, etc.), and Arthropoda (arthropods). Vertebrates, which are a subphylum of the phylum Chordata, include such familiar animals as fish, amphibians, reptiles, birds, and mammals.

Morphography includes the systematic exploration and tabulation of the facts involved in the recognition of all the recent and extinct kinds of animals and their distribution in space and time. The museum-makers of olden times and their modern representatives, the curators and describers of zoological collections; the early explorers and the modern naturalists and writers on zoogeography; and the collectors of fossils and present-day paleontologists are the chief varieties of zoological workers coming under this heading. Gradually, since the time of John Hunter and Georges Cuvier, anatomical study has associated itself with the more superficial morphography until today no one considers a study of animal form of much value if it does not include internal structure, histology, and embryology in its scope.

Botany

Botany is the branch of biology that deals with the study of plants including their structure, properties, and biochemical processes. Also included are plant classification and the study of plant diseases and of interactions with the environment. The principles and findings of botany have provided the base for such applied sciences as agriculture, horticulture, and forestry.

Plants were of paramount importance to early man; he depended upon them as sources of food, shelter, clothing, medicine, ornament, tools, and magic. Today it is known that, in addition to their practical and economic values, green plants are indispensable to all life on Earth: through the process of photosynthesis, plants transform energy from the sun into the chemical energy of food, which makes all life possible. A second unique and important capacity of green plants is the formation and release of oxygen as a by-product of photosynthesis. The oxygen of the atmosphere, so absolutely essential to many forms of life, represents the accumulation of over 3,500,000,000 years of photosynthesis by green plants.

Although the many steps in the process of photosynthesis have become fully understood only in recent years, even in prehistoric times man somehow recognized intuitively that some important relation existed between the sun and plants. Such recognition is suggested by the fact that, in primitive tribes and early civilizations, worship of the sun was often combined with the worship of plants.

Earliest man, like the other anthropoid mammals (e.g., apes, monkeys), depended totally upon the natural resources of his environment, which, until he developed methods for hunting, consisted almost completely of plants. The behaviour of pre-Stone Age man can be inferred by studying the botany of aboriginal peoples in various parts of the world. Isolated tribal groups in South America, Africa, and New Guinea, for example, have extensive knowledge about plants and distinguish hundreds of kinds according to their utility, as edible, poisonous, or otherwise important in their culture. They have developed surprisingly sophisticated systems of nomenclature and classification, which approximate the binomial system (i.e., generic and specific names) found in modern biology. The urge to recognize different kinds of plants and to give them names thus seems to be as old as the human race.

In time plants were not only collected by primitive man but also grown by him. This domestication resulted not only in the development of agriculture but also in a greater stability of human populations that had previously been nomadic. From the settling down of agricultural peoples in places where they could depend upon adequate food supplies came the first villages and the earliest civilizations.

Because of the long preoccupation of man with plants, a large body of folklore, general information, and actual scientific data has accumulated, which has become the basis for the science of botany.

Divisions of Botany

Botany today is such a large and varied area and has many important uses in a wide variety of disciplines, that it has evolved several niches and offshoots. The most notable are as follows, but this list is not exhaustive:

Plant Pathology

It is often tragedy that drives scientific discovery and the Irish Potato Blight of the 19th century is the one period where most advances in research of plant diseases were made - many researchers put their efforts into plant pathology - the reasons should be obvious, if disease wipes out a large proportion of a staple crop then people starve. The Irish 1840s "Potato Famine" drove immigration for a large number of Irish to North America largely because of the disease that had affected the crops they had come to rely on. With Potato Blight, the tuber or the roots rot, rendering the whole crop useless. The mass spread of this disease caused one of the greatest famines anywhere in the western world and has never been seen since. Today, plant pathology is still a vital component of botany as we attempt to adapt crops to the changing climate and keep one step ahead of evolving diseases.

An ongoing research problem for plant pathologists is how to tackle the many diseases that threaten bananas. The limited genetic make-up and the seedless nature of modern cultivars means that several types have already gone extinct and several others remain under threat. Plant pathology is vital to understanding disease as well as the potential problems caused by climate change.

Plant Ecology

Plant ecology differs from botany in that it is more concerned with how plants interact with their environment - with and in soils, with animal species, how it reacts to ecological change such as climate change and many more issues of the wider landscape .This is vital in a world understanding and adapting to climate change in working out how to breed or otherwise alter our staple plants to survive and adapt to changing environments. It is also vital to the growing science of GM crops and cross-breeding for hardier crops to grow in what we call "marginal landscapes" as human population increases and require more land.

Plant ecology recognises eleven major types of environment in which plant life exists: tropical forest, temperate forest, coniferous forests, tropical savannah, temperate grassland/plain, desert and other arid ecosystems, Mediterranean-like regions, terrestrial wetlands, freshwater ecology, coastal/marine ecology and tundra. Each has its own ecological profile, and balanced plant and animal life and how these interact is as important as each plant's profile in order that we can understand their evolution.

Palaeobotany

Palaeobotany is the study of extinct plants or fossilised plants recovered from geological strata. Palaeobotanists will also study fossilised algae, bacteria, fungi and lichens - just as botanists study modern plants in these taxa. Palaeobotany has been fundamental to understand the changing climate of the past. It shows us what the ecological make-up of each climate zone was like at any given period of our distant past. Extensive studies in South America have shown precisely how and when the tropical rainforests there developed and what conditions led to their evolution. We also expect that such information will tell us about how they might change in future in light of a changing climate. Palaeobotany is also shedding light on the nature and the extent of plant species during the Ice Age. One study examined the land bridge between modern Siberia and Alaska, showing how people lived there and may have crossed this land bridge in antiquity.

Archaeobotany

Though most researchers in this area come from an archaeological or anthropological background, a botanist will have the tools to enter into the field of studying how people in the past used plants. This discipline can be functional - in terms of looking at the spread of crop farming, wetland drainage, irrigation and other forms of ecological engineering - but it can also teach us about how people of the past experienced plants. Some plants have medicinal properties and others have had - and continue to have - spiritual significance. Sometimes understanding the plant can also help us to understand the beliefs of the past. Archaeobotanists will not just look at plant remains in the soil, but residues in broken pottery and on stone tools for example.

In understanding how environments change, it is vital to separate changes wrought upon a landscape by nature from changes made as a result of human ecological engineering. Seeds and other plant remains can be great indicators for change, and archaeobotany has been vital for tracking the spread of the Neolithic Revolution as landscapes undergo wholesale change purely because of human changes - and by extension, in some cases the spread of the market economy.

Forensic Botany

Criminal investigations use whatever evidence they can find and in the last few decades, forensics has been the area with the most innovation and development. What can plants tell us about a crime scene, or a body? As with pollen and other minute plant remains, when we understand the make-up and profile of certain plants, plant evidence can tell us quite a lot.

A forensic botanist will not typically study human remains but will look at plant evidence recovered from bodies or will use criteria that could tell us where a body has been or is presently buried. One of the most important cases ever, and criminal historians will say that it was the case where botany was first employed in a criminal trial, is that of the abduction of Charles Lindbergh's son. A tree expert was called as a witness for the prosecution and his evidence indicated that the abductor was most likely an expert carpenter to have designed and built such a complex and unique ladder to aid the abduction; the strong testimony was permitted into the evidence base. Forensic botany, and the use of plants as evidence in a criminal trial, has come a long way since then and evidence has been known to determine whether death was a result of homicide, suicide or misadventure.

Methods in Botany

Morphological Aspects

The invention of the compound microscope provided a valuable and durable instrument for the investigation of the inner structure of plants. Early plant morphologists, especially those studying cell structure, were handicapped as much by the lack of adequate knowledge of how to prepare specimens as they were by the imperfect microscopes of the time. A revolution in the effectiveness of microscopy occurred in the second half of the 19th century with the introduction of techniques for fixing cells and for staining their component parts. Before the development of these techniques, the cell, viewed with the microscope, appeared as a minute container with a dense portion called the nucleus. The discovery that parts of the cell respond to certain stains made observation easier. The development of techniques for preparing tissues of plants for microscopic examination was continued in the 1870s and 1880s and resulted in the gradual refinement of the field of nuclear cytology, or karyology. Chromosomes were recognized as constant structures in the life cycle of cells, and the nature and meaning of meiosis, a type of cell division in which the daughter cells have half the number of chromosomes of the parent, was discovered; without this discovery, the significance of Mendel's laws of heredity might have gone unrecognized. Vital stains, dyes that can be used on living material, were first used in 1886 and have been greatly refined since then.

Improvement of the methodology of morphology has not been particularly rapid, even though satisfactory techniques for histology, anatomy, and cytology have been developed. The embedding of material in paraffin wax, the development of the rotary microtome for slicing very thin sections of tissue for microscope viewing, and the development of stain techniques are refinements of previously known methods. The invention of the phase microscope made possible the study of unfixed and unstained living material—hopefully nearer its natural state. The development of the electron microscope, however, has provided the plant morphologist with a new dimension of magnification of the structure of plant cells and tissues. The fine structure of the cell and of its components, such as mitochondria and the Golgi apparatus, have come under intensive study. Knowledge of the fine structure of plant cells has enabled investigators to determine the sites of important biochemical activities, especially those involved in the transfer of energy during photosynthesis and respiration. The scanning electron microscope, a relatively recent development, provides a three-dimensional image of surface structures at very great magnifications.

For experimental research on the morphogenesis of plants, isolated organs in their embryonic stage, clumps of cells, or even individual cells are grown. One of the most interesting techniques developed thus far permits the growing of plant tissue of higher plants as single cells; aeration and continuous agitation keep the cells suspended in the liquid culture medium.

Physiological Aspects

Plant physiology and plant biochemistry are the most technical areas of botany; most major advances in physiology also reflect the development of either a new technique or the dramatic refinement of an earlier one to give a new degree of precision. Fortunately, the methodology of measurement has been vastly improved in recent decades, largely through the development of various electronic devices. The phytotron at the California Institute of Technology represents the first serious attempt to control the environment of living plants on a relatively large scale; much important information has been gained concerning the effects on plants of day length and night length and the effects on growth, flowering, and fruiting of varying night temperatures. Critical measurements of other plant functions have also been obtained.

Certain complex biochemical processes, such as photosynthesis and respiration, have been studied stepwise by immobilizing the process through the use of extreme cold or biochemical inhibitors and by analyzing the enzymatic activity of specific cell contents after spinning cells at very high speeds in a centrifuge. The pathways of energy transfer from molecule to molecule during photosynthesis and respiration have been determined by biophysical methods, especially those utilizing radioactive isotopes.

An investigation of the natural metabolic products of plants requires, in general, certain standard biochemical techniques—e.g., gas and paper chromatography, electrophoresis, and various kinds of spectroscopy, including infrared, ultraviolet, and nuclear magnetic resonance. Useful information on the structure of the extremely large cellulose molecule has been provided by X-ray crystallography.

Ecological Aspects

When plant ecology first emerged as a subscience of botany, it was largely descriptive; today, however, it has become a common meeting ground for all the plant sciences, as well as for other sciences. In addition, it has become much more quantitative. As a result, the tools and methods of plant ecologists are those available for measuring the intensity of the environmental factors that impinge on the plant and the reaction of the plant to these factors. The extent of the variability of many physical factors must be measured. The integration and reporting of such measurements, which cannot be regarded as constant, may therefore conceal some of the most dynamic and significant aspects of the environment and the responses of the plant to them. Because the physical environment is a complex of biological and physical components, it is measured by biophysical tools. The development of electronic measuring and recording devices has been crucial for a better understanding of the dynamics of the environment. Such devices, however, produce so much information that computer techniques must be used to reduce the data to meaningful results.

The ecologist, concerned primarily with measuring the effect of the external environment on a plant, adapts the methodology of the plant physiologist to field conditions.

The plant sociologist, on the other hand, is concerned with both the relation of different kinds of plants to each other and the nature and constitution of their association into natural communities. One widely used technique in this respect is to count the various kinds of plants within a standard area in order to determine such factors as the percentage of ground cover, dominance of species, aggressiveness, and other characteristics of the community. In general, the plant sociologist has relatively few quantitative factors to measure and must therefore take a subjective and intuitive approach, which, nevertheless, gives extremely useful results and some degree of predictability.

Some ecologists are most concerned with the inner environment of the plant and the way in which it reacts to the external environment. This approach, which is essentially physiological and biochemical, is useful for determining energy flow in ecosystems. The physiological ecologist is also concerned with evaluating the adaptations that certain plants have made toward survival in a hostile environment.

The techniques and methodology of plant ecology are as diverse and as varied as the large number of sciences that are drawn upon by ecologists. Completely new techniques, although few, are important; among them are techniques for measuring the amount of radioactive carbon-14 in plant deposits up to 50,000 years old. The most important new method in plant ecology is the rapidly growing use of computer techniques for handling vast amounts of data. Furthermore, modern digital computers can be used to simulate simple ecosystems and to analyze real ones.

Taxonomic Aspects

Experimental research under controlled conditions, made possible by botanical gardens and their ranges of greenhouses and controlled environmental chambers, has become an integral part of the methodology of modern plant taxonomy.

A second major tool of the taxonomist is the herbarium, a reference collection consisting of carefully selected and dried plants attached to paper sheets of a standard size and filed in a systematic way so that they may be easily retrieved for examination. Each specimen is a reference point representing the features of one plant of a certain species; it lasts indefinitely if properly cared for, and, if the species becomes extinct in nature—as hundreds have—it remains the only record of the plant's former existence. The library is also an essential reference resource for descriptions and illustrations of plants that may not be represented in a particular herbarium.

One of the earliest methods of the taxonomist, the study of living plants in the field, has benefitted greatly by fast and easy methods of transportation; botanists may carry on fieldwork in any part of the world and make detailed studies of the exact environmental conditions under which each species grows.

During the present century, many new approaches have been applied to the elucidation of problems in systematic botany. The transmission electron microscope and the scanning electron microscope have added to the knowledge of plant morphology, upon which classical taxonomy so much depends.

Refined methods for cytological and genetical studies of plants have given the taxonomist new insights into the origin of the great diversity among plants, especially the mechanisms by which new species arise and by which they then maintain their individuality in nature. From such studies have arisen further methods and also the subdisciplines of cytotaxonomy, cytogenetics, and population genetics.

Phytochemistry, or the chemistry of plants, one of the early subdivisions of organic chemistry, has been of great importance in the identification of plant substances of medicinal importance. With the development of new phytochemical methods, new information has become available for use in conjunction with plant taxonomy; thus has arisen the modern field of chemotaxonomy, or biochemical systematics. Each species tends to differ to some degree from every other species, even in the same genus, in the biochemistry of its natural metabolic products. Sometimes the difference is subtle and difficult to determine; sometimes it is obvious and easily perceptible. With new analytical techniques, a large number of individual compounds from one plant can be identified quickly and with certainty. Such information is extremely useful in adding confirmatory or supplemental evidence of an objective and quantitative nature. An interesting by-product of chemical plant taxonomy has resulted in understanding better the restriction of certain insects to specific plants.

Computer techniques have recently been applied to plant taxonomy to develop a new field, numerical taxonomy, or taximetrics, by which relationships between plant species or those within groups of species are determined quantitatively and depicted graphically. Another method measures the degree of molecular similarity of deoxyribonucleic acid (DNA) molecules in different plants. By this procedure it should be possible to determine the natural taxonomic relationships among different plants and plant groups by determining the extent of the relationship of their DNA's: closely related plants will have more similarities in their DNA's than will unrelated ones.

Scope and Importance

Study of plants is equally important because it covers almost all animal life on Earth by generating a large proportion of oxygen that provides us and other organisms with aerobic respiration with the chemical energy that is essential to be alive. Cyanobacteria and Algae are the major groups of organisms that perform photosynthesis, which is a process that consumes the energy of sunlight to convert carbon dioxide and water into sugar that can be utilized both as a source of chemical energy and organic molecules that exist in structural components of cells. Plants liberate oxygen into the atmosphere as a by-product of photosynthesis, which is needed by all living things for cellular respiration. Also, they influence the global carbon, water cycles, and plant roots stabilize soils, that prevents soil erosion.

All living things were organized as plants or animals previously, and botany encompasses the study of all organisms and does not include animals. Botanists test both the internal functions and processes within the organelles of plants, tissues and plants communities. At every level, botanists are concerned with the classification, evolution, phylogeny or function of plant life.

The most precise definition of 'plant' involves land plants or embryophytes that include seed plants such as pines and flowering plants, and the free-sporing cryptogams involve mosses, ferns, club-mosses, and hornworts. Embryophytes are multicellular eukaryotes originating from an ancestor that gains its energy from sunlight through photosynthesis. Their life cycles are with alternating haploid and diploid phases. The embryophytes phase of sexual haploid is called gametophyte. It nourishes the developing diploid embryo sporophytes in the tissues in the seed plants, where gametophyte is nourished by its parent sporophyte. Other groups of organisms that were studied by botanists involve bacteria, fungi(including lichen generating fungi, and non-chlorophyte algae) and viruses. However, stress is still given to these fungi and photosynthetic protists that are usually covered in introductory courses.

Plant Biochemistry

Plant biochemistry is defined as the study of chemical process used by plants. Some of these processes are utilized in their primary metabolisms such as photosynthetic Calvin cycle and crassulacean acid metabolism. Others produce special substances like lignin and cellulose that is utilized to build their bodies and secondary products such as aroma compounds.

Various other groups of photosynthetic eukaryotes are collectively called as algae that have special organelles such as chloroplasts. Chloroplasts are originated from cyanobacteria that produced endosymbiotic relationships with ancient plants. Cyanobacteria and Chloroplasts consist of the blue-green pigment called chlorophyll a. It absorbs light in the orange-red and blue-violet parts of the spectrum while reflecting and transmitting the green light that view as the characteristic color of these organisms.

Plants, Climate, and Environmental Change

The plant responses towards the climate and environmental changes indicate how these changes affect the functions of ecosystem and productivity. For example, plant phenology can be an important proxy for temperature, and the biological impact of climate change. The analysis of fossil resides in the sediments from thousands of years ago allows the reconstruction of past climates. Ozone depletion can expose plants to high ultraviolet rays that result in lower growth rates. Moreover, knowledge of community ecology and taxonomy is vital to understand habitat destruction and species extinction.

There are 'n' types of plants available to botanists since this field is so broad. Botanists that are interested in the ecology of plant interactions with other organisms and environment. Botanists of other fields search to find new species or perform experiments to discover the growth of plants under varied condition. Some use microscopes to study detailed structure of every cell. Botanists study processes that take place on a time scale ranging from a fraction of second in individual cells to those that uncover eons of evolutionary time.

Marine Biology

Marine biology is the study of marine organisms, their behaviors and interactions with the environment. Marine biologists study biological oceanography and the associated fields of chemical, physical, and geological oceanography to understand marine organisms. Marine biology is a very broad area, so most researchers select a particular area of interest and specialize in it. Specializations can be based on a particular species, group, behavior, technique, or ecosystem. Molecular biology is a related area of specialization in marine biology. Researchers apply molecular techniques to many environments ranging from coastal marshes to the deep sea and to various organisms such as viruses, plants, and fish.

Like all scientific disciplines, the study of marine biology also follows the scientific method. The overriding goal in all of science is to find the truth. Although following the scientific method is not by any means a rigid process, research is usually conducted systematically and logically to narrow the inevitable margin of error that exists in any scientific study, and to avoid as much bias on behalf of the researcher as possible. The primary component of scientific research is characterization by observations. Hypotheses are then formulated and then tested based on a number of observations in order to determine the degree to which the hypothesis is a true statement and whether or not it can be accepted or rejected. Testing is then often done by experiments if hypotheses can produce predictions based on the initial observations.

The essential elements of the scientific method are iterations and recursions of the following four steps:

- Characterization (observation)
- Hypothesis (a theoretical, hypothetical explanation)

- Prediction (logical deduction from the hypothesis)
- Experiment (test of all of the above)

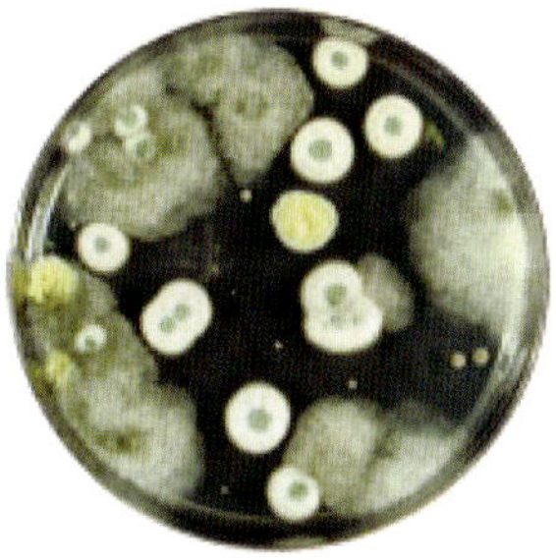

These steps are all used in the study of marine biology, which includes numerous sub fields including:

- Microbiology: The study of microorganisms, such as bacteria, viruses, protozoa and algae, is conducted for numerous reasons. One example is to understand what role microorganisms play in marine ecosystems. For example, bacteria are critical to the biological processes of the ocean, as they comprise 98% of the ocean's biomass, which is the total weight of all organisms in a given volume. Microbiology is also important to our understanding of the food chain that connects plants to herbivorous and carnivorous animals. The first level in the food chain is primary production, which occurs at the microbial level. This is an important biological activity to understand as primary production drives the entire food chain.

  Scientists also study marine microbiology to find new organisms that may be used to help develop medicines and find cures for diseases and other health problems.

- Fisheries and Aquaculture: to protect biodiversity and to create sustainable seafood sources because of the world's dependence on fish for protein. There are many areas of study in this field.
  - The ecology of fisheries includes the study of their population dynamics, reproduction, behavior, food webs, and habitat.
  - Fisheries management includes studies on the impact of overfishing, habitat destruction, pollution and toxin levels, and ways to increase populations for sustainability as seafood.

  - Aquaculture includes research on the development of individual organisms and their environment. The objective is most often to develop the knowledge needed to cultivate certain species in a designated area in open water or in captivity in order to meet consumer demand. Technological advances have enabled seafood "farms" to produce high-demand products that traditional commercial fisheries cannot meet. This is a controversial area however, and an issue that will become of greater importance as our fish stocks continue to decline.

- Environmental marine biology: includes the study of ocean health. It is important for scientists to determine the quality of the marine environment to ensure water quality is sufficient to sustain a healthy environment. Coastal environmental health is an important area of environmental marine biology so that scientists can determine the impact of coastal development on water quality for the safety of people visiting the beaches and to maintain a healthy marine environment. Pollutants, sediment, and runoff are all potential threats to marine health in coastal areas. Offshore marine environmental health is also studied. For example, an environmental biologist might be required to study the impact of an oil spill or other chemical hazard in the ocean. Environmental biologists also study Benthic environments on the ocean bottom in order to understand such issues as the chemical makeup of sediment, impact of erosion, and the impact of dredging ocean bottoms on the marine environment.

- Deep-sea ecology: advances in technology of equipment needed to explore the deep sea have opened the door to the study of this largely unknown space in the sea. The biological characteristics and processes in the deep-sea environment are of great interest to scientists. Research includes the study of deep ocean gases as an alternate energy source, how animals of the deep live in the dark, cold, high pressure environment, deep sea hydrothermal vents and the lush biological communities they support.
- Ichthyology: is the study of fishes, both salt and freshwater species. There are some 25,000+ species of fishes including: bony fishes, cartilaginous fishes, sharks, skates, rays, and jawless fishes. Ichthyologists study all aspects of fish from their classification, to their morphology, evolution, behavior, diversity, and ecology. Many ichthyologists are also involved in the field of aquaculture and fisheries.

- Marine mammology: This is the field of interest to most aspiring marine biologists. It is the study of cetaceans—families of whales and dolphins, and pinnipeds (seals, sea lions, and the walrus). Their behaviors, habitats, health, reproduction, and populations are all studied. These are some of the most fascinating creatures in the sea; therefore, this is an extremely competitive field, and difficult to break into because the competition for research funding is also quite heavy.

One area of research currently being conducted on whales is the impact of military sonar on their health and well-being. The scientific community believes that high frequency sound waves cause internal damage and bleeding in the brains of whales, yet the military denies this claim. Military sonar can also interfere with the animal's own use of sonar for communication and echolocation. More research is needed; however, in recent years science has proven the claims to be valid and the military has begun limiting its use of sonar in specific areas.

- Marine ethology: The behavior of marine animals is studied so that we understand the animals that share the planet with us. This is also an important field for help in understanding how to protect endangered species, or how to help species whose habitats are threatened by man or natural phenomena. The study of marine animal behavior usually falls under the category of ethology because most often marine species must be observed in their natural environment, although there are many marine species observed in controlled environments as well. Sharks are most often studied in their natural habitat for obvious reasons.

Marine Life

The phrase 'marine life' refers to organisms that live in salt water. These can include a diverse array of plants, animals and microbes (tiny organisms) such as bacteria and archaea.

Marine Life Are Adapted to Life in Saltwater

From the perspective of a land animal like us, the ocean can be a harsh environment. However, marine life are adapted to live in the ocean. Characteristics that help marine life thrive in a saltwater environment include the ability to regulate their salt intake or deal with large quantities of salt water, adaptations to obtain oxygen (e.g., a fish's gills), being able to withstand high water pressures, living in a place where they can get enough light, or being able to adjust to a lack of light. Animals and plants that live on the edge of the ocean, such as tide pool animals and plants, also need to deal with extremes in water temperature, sunlight, wind and waves.

Types of Marine Life

There is a huge diversity in marine species. Marine life can range from tiny, single-celled organisms to gigantic blue whales, which are the largest creatures on Earth. Below is a list of the major phyla, or taxonomic groups, of marine life.

Major Marine Phyla

The classification of marine organisms is always in flux. As scientists discover new species, learn more about the genetic makeup of organisms, and study museum specimens, they debate how organisms should be grouped. More information about the major groups of marine animals and plants is listed below.

Marine Animal Phyla

Some of the most well-known marine phyla are listed below.

- Annelida - this phylum contains segmented worms. An example of a segmented marine worm is the Christmas tree worm.
- Arthropoda - Arthropods have a segmented body, jointed legs and a hard exoskeleton for protection. This group includes lobsters and crabs.
- Chordata - Humans are in this phylum, which also includes marine mammals (cetaceans, pinnipeds, sirenians, sea otters, polar bears), fish, tunicates, seabirds and reptiles.
- Cnidaria - This is a diverse phylum of animals, many of whom have stinging structures called nematocysts. Animals in this phylum include corals, jellyfish, sea anemones, sea pens and hydras.
- Ctenophora - These are jelly-like animals, such as comb jellies, but they don't have stinging cells.
- Echinodermata - This is one of my favorite phylums. It includes such beautiful animals as sea stars, brittle stars, basket stars, sand dollars and sea urchins.
- Mollusca - This phylum includes snails, sea slugs, octopuses, squids, and bivalves such as clams, mussels and oysters.
- Porifera - This phylum includes sponges, which are living animals. They can be very colorful and come in a diverse array of shapes and sizes.

Marine Plant Phyla

There are also several phyla of marine plants. These include the Chlorophyta, or green algae, and the Rhodophyta, or red algae.

Marine Habitats and Ecosystems

Marine ecosystems are part of the earth's aquatic ecosystem. The habitats that make up this vast system range from the productive nearshore regions to the barren ocean floor. The marine waters may be fully saline, brackish or nearly fresh. The saline waters have a salinity of 35-50 ppt (= parts per thousand). The freshwater has a salinity of less than 0.5 ppt. The brackish water lies in between these 2. Marine habitats are situated from the coasts, over the continental shelf to the open ocean and deep sea. The ecosystems are sometimes linked with each other and are sometimes replacing each other in other geographical regions. The reason why habitats differ from another is because of the physical factors that influence the functioning and diversity of the habitats. These factors are temperature, salinity, tides, currents, wind, wave action, light and substrate.

Marine ecosystems are home to a host of different species ranging from planktonic organisms that form the base of the marine food web to large marine mammals. Many species rely on marine ecosystems for both food and shelter from predators. They are very important to the overall health of both marine and terrestrial environments. Coastal habitats are those above the spring high tide limit or above the mean water level in non-tidal waters. They are close to the sea and include habitats such as coastal dunes and sandy shores, beaches , cliffs and supralittoral habitats. Coastal habitats alone account for approximately 30% of all marine biological productivity. The diversity and productivity are also important for humans. These habitats provide a rich source of food and income. They also support species that serve as animal feed, fertilizers, additives in food and cosmetics. Habitats such as mangroves and seagrasses protect the coastlines from wave action and erosion. Other areas provide sediment sinks or act as filtering systems.

Despite the importance of marine ecosystems, increased human activities have caused significant damage or are serious threats to the marine biodiversity. These activities can be overfishing, pollution, introduction of exotic species or coastal development. For this reason, conservation plans are necessary to save the marine ecosystems from being lost.

Molecular Biology

Molecular biology is the study of living things at the level of the molecules which control them and make them up. While traditional biology concentrated on studying whole living organisms and how they interact within populations, molecular biology strives to understand living things by examining the components that make them up. Both approaches to biology are equally valid, although improvements to technology have permitted scientists to concentrate more on the molecules of life in recent years.

Molecular biology is a specialised branch of biochemistry, the study of the chemistry of molecules which are specifically connected to living processes. Of particular importance to molecular biology are the nucleic acids (DNA and RNA) and the proteins which are constructed using the genetic instructions encoded in those molecules. Other biomolecules, such as carbohydrates and lipids may also be studied for the interactions they have with nucleic acids and proteins. Molecular biology is often separated from the field of cell biology, which concentrates on cellular structures (organelles and the like), molecular pathways within cells and cell life cycles.

The molecules which form the basis of life provide scientists with a more predictable and mechanistic tool for scientists to study. Working with whole organisms (or even just whole cells) can be unpredictable, with the outcome of experiments relying on the interaction of thousands of molecular pathways and external factors. Molecular biology provides scientists with a toolkit with which they may "tinker" with the way life works. They may use them to determine the function of single genes or proteins, and find out what would happen if that gene or protein was absent or faulty. Molecular biology is used to examine when and why certain genes are switched "on" or "off". An understanding of each of the factors has granted scientists a deeper understanding of how living things work, and used this knowledge to develop treatments for when living things don't work so well.

Subdisciplines of Molecular Biology

Comparative Genomics

This is the study of human genetics by comparisons with model organisms such as mice, the fruit fly and the bacterium E. coli.

DNA Forensics

This is the use of DNA for identification. Some examples of DNA use are to establish paternity in child support cases, to establish the presence of a suspect at a crime scene and to identify accident victims.

Functional Genomics

This is the study of genes, their resulting proteins and the role played by the proteins in the body's biochemical processes.

Gene Therapy

This is an experimental procedure aimed at replacing, manipulating or supplementing nonfunctioning or misfunctioning genes with healthy genes.

Genomics

This is the study of genes and their functions.

Molecular Genetics

This is the study of macromolecules important in biological inheritance.

Pharmacogenomics

This is the study of the interaction of an individual's genetic makeup and response to a drug.

Proteomics

This is the study of the full set of proteins encoded by a genome.

Structural Genomics

This is the effort to determine the three-dimensional structures of large numbers of proteins using both experimental techniques and computer simulation.

Toxicogenomics

This is the study of how genomes respond to environmental stressors or toxicants.

Concepts in Molecular Biology

Key concepts in molecular biology are mechanism, information, and gene. Hence, major tasks for philosophers of molecular biology have been and continue to be analyzing the concepts of mechanism, information, and gene in order to understand how they have been, are, and should be used.

1. Mechanism

Molecular biologists discover and explain by identifying and elucidating mechanisms, such as DNA replication, protein synthesis, and the myriad mechanisms of gene expression. The phrase "theory of molecular biology" was not used above and for good reason; general knowledge in the field is represented by diagrams of mechanisms. Discovering the mechanism that produces a phenomenon is an important accomplishment for several reasons. First, knowledge of a mechanism shows how something works: elucidated mechanisms provide understanding. Second, knowing how a mechanism works allows predictions to be made based upon the regularity in mechanisms. For example, knowing how the mechanism of DNA base pairing works in one species allows one to make predictions about how it works in other species, even if conditions or inputs are changed. Third, knowledge of mechanisms potentially allows one to intervene to change what the mechanism produces, to manipulate its parts to construct experimental tools, or to repair a broken, diseased mechanism. In short, knowledge of elucidated mechanisms provides understanding, prediction, and control. Given the general importance of mechanisms and the fact that mechanisms play such a central role in the field of molecular biology, it is not surprising that philosophers of biology pioneered analyzing the concept of mechanism.

Starting in the 1990s, a number of philosophers focused squarely on how the concept of a mechanism functions in science generally and molecular biology specifically. A number of characterizations of what a mechanism is have emerged over the years. Phyllis McKay Illari and Jon Williamson have more recently offered a characterization that draws on the essential features of all the earlier contributions.

A mechanism for a phenomenon consists of entities and activities organized in such a way that they are responsible for the phenomenon.

As an example, consider the phenomenon of DNA replication. As Watson and Crick famously noted upon discovery of the structure of DNA, the macromolecule's structure pointed to the mechanism of DNA replication.

It has not escaped our notice that the specific pairing we have postulated immediately suggests a possible copying mechanism for the genetic material.

In short, the double helix of DNA (an entity with an organization) unwinds (an activity) and new component parts (entities) bond (an activity) to both parts of the unwound DNA helix. DNA is a nucleic acid composed of several subparts: a sugar-phosphate backbone and nucleic acid bases. When DNA unwinds, the bases exhibit weak charges, properties that result from slight asymmetries in the molecules. These weak charges allow a DNA base and its complement to engage in the activity of forming hydrogen (weak polar) chemical bonds; the specificity of this activity is due to the topological arrangements of the weak polar charges in the subparts of the base. Ultimately, entities with polar charges enable the activity of hydrogen bond formation. After the complementary bases align, then the backbone forms via stronger covalent bonding. The mechanism proceeds with unwinding and bonding together (activities) new parts, to produce two helices (newly formed entities) that are (more or less faithfully) copies of the parent helix.

Scientists rarely depict all the particular details when describing a mechanism; representations are usually schematic. Such representations may be called a "model of a mechanism", or "mechanism schema". A mechanism schema is a truncated abstract description of a mechanism that can be instantiated by filling it with more specific descriptions of component entities and activities. An example is James Watson's diagram of his version of the central dogma of molecular biology:

DNA → RNA → protein.

This is a schematic representation (with a high degree of abstraction) of the mechanism of protein synthesis, which can be instantiated with details of DNA base sequence, complementary RNA sequence, and the corresponding order of amino acids in the protein produced by the more specific mechanism. Molecular biology textbooks are replete with diagrams of mechanism schemas. A mechanism schema can be instantiated to yield a description of a particular mechanism. In contrast, a mechanism sketch cannot (yet) be instantiated; components are (as yet) unknown. Sketches have black boxes for missing components or grey boxes whose function is known but whose entities and activities that carry out that function are not yet elucidated.

2. Information

The language of information is used ubiquitously by molecular biologists. Genes as linear DNA sequences of bases are said to carry "information" for the production of proteins. During protein synthesis, the information is "transcribed" from DNA to messenger RNA and then "translated" from RNA to protein. With respect to inheritance, it is often said that what is passed from one generation to the next is the "information" in the genes, namely the linear ordering of bases along complementary DNA strands. Historians of biology have tracked the entrenchment of information-talk in molecular biology since its introduction.

The question for philosophers of biology is whether an analysis of the concept of information can capture the various ways in which the concept is used in molecular biology. The usage of "information" in the mathematical theory of communication is too impoverished to capture the molecular biological usage, since the coded sequences in the DNA are more than just a signal with some number of bits that may or may not be accurately transmitted. Conversely, the usage in cognitive neuroscience, with its talk of "representations" may be said to be too rich, since the coded sequences in the DNA are also not said to have within them a representation of the structure of the protein. No definition of "information" as it is used in molecular biology has yet received wide support among philosophers of biology.

Stephen Downes helpfully distinguishes three positions on the relation between information and the natural world:

1. Information is present in DNA and other nucleotide sequences. Other cellular mechanisms contain no information.
2. Information is present in DNA, in other nucleotide sequences and other cellular mechanisms, for example cytoplasmic or extra-cellular proteins; and in many other media, for example, the embryonic environment or components of an organism's wider environment.
3. DNA and other nucleotide sequences do not contain information, nor do any other cellular mechanisms.

These options may be read either ontologically or heuristically. A heuristic reading of, for instance, views the talk of information in molecular biology as useful in providing a way of talking and in guiding research. And so the heuristic benefit of the information concept can be defended without making any commitment to the ontological status. Indeed, one might argue that a vague and open-ended use of information is valuable for heuristic purposes, especially during early discovery phases in the development of a field.

Philosophers' discussions of the concept of information in biology have also focused on its ontological reading. Three different philosophical accounts of information serve as exemplars of Downes' three categories. Ulrich Stegmann provides an example of Downes' first category with his analysis of template-directed synthesis. (Stegmann does explicitly allow that components other than nucleotide sequences might contain what he calls instructional information. However, his only example is a thought experiment involving enzymes linearly ordered along a membrane; nothing of the sort is known to actually exist or even seems very likely to exist.) Stegmann calls this the sequentialization view. Stegmann's instructional account of genetic information requires that the component carrying the information satisfy the following conditions: an advance specification of the kind and order of steps that yield a determinate outcome if the steps are carried out. On his account, DNA qualifies as an instructional information carrier for replication, transcription and translation. The sequence of bases provides the order. The hydrogen bonding between specific bases and the genetic code provide the specific kinds of steps. And the mechanisms of replication, transcription, and translation yield certain outcomes: a copy of the DNA double helix, an mRNA, and a linear order of amino acids. Also, because DNA carries information for a specific outcome, an error can occur as the mechanism operates to produce that outcome; hence Stegmann's account allows for errors and error-correcting mechanisms (such as proof reading mechanisms that correct DNA mutations).

Eva Jablonka is an example of Downes' second category. She argues that information is ubiquitous. She defines information as follows: a source becomes an informational input when an interpreting receiver can react to the form of the source (and variations in this form) in a functional manner. She claims a broad applicability of this definition. The definition, she says, accommodates information stemming from environmental cues as well as from evolved signals, and calls for a comparison between information-transmission in different types of inheritance systems — the genetic, the epigenetic, the behavioral, and the cultural-symbolic. On this view, genes have no theoretically privileged informational status.

In line with Downes' third category, C. Kenneth Waters argues that information is a useful term in rhetorical contexts, such as seeking funding for DNA sequencing by claiming that DNA carries information. However, from an ontological perspective, Waters claims that explication of DNA's causal role has no need for the concept of information. Genes, he argues, should not be viewed as "immaterial units of information". Waters' focus is on stretches of DNA whose causal roles are as actual specific difference makers in genetic mechanisms. Talk of information is not needed; causal role function talk is sufficient.

3. Gene

The question of whether classical, Mendelian genetics could be reduced to molecular biology motivated philosophers to consider the connectability of the term they shared: the gene. Investigations of reduction and scientific change raised the question of how the concept of the gene evolved over time, figuring prominently in C. Kenneth Waters', Philip Kitcher's and Raphael Falk's work. Over time, however, philosophical discussions of the gene concept took on a life of their own, as philosophers raised questions independent of the reduction debate: What is a gene? And, is there anything causally distinct about DNA?

Falk explicitly asked philosophers and historians of biology, "What is a Gene?" Discoveries such as overlapping genes, split genes, and alternative splicing made it clear that simply equating a gene with an uninterrupted stretch of DNA would no longer capture the complicated molecular-developmental details of mechanisms such as gene expression. In an effort to answer Falk's question, two general trends have emerged in the philosophical literature: first, distinguish multiple gene concepts to capture the complex structural and functional features separately, or second, rethink a unified gene concept to incorporate such complexity.

A paradigmatic example of the first line came from Lenny Moss's distinction between Gene-P and Gene-D. Gene-P embraced an instrumental preformationism (providing the "P"); it was defined by its relationship to a phenotype. In contrast, Gene-D referred to a developmental resource (providing the "D"); it was defined by its molecular sequence. An example will help to distinguish the two: When one talked about the gene for cystic fibrosis, the most common genetic disease affecting populations of Western European descent, the Gene-P concept was being utilized; the concept referred to the ability to track the transmission of this gene from generation to generation as an instrumental predictor of cystic fibrosis, without being contingent on knowing the causal pathway between the particular sequence of DNA and the ultimate phenotypic disease. The Gene-D concept, in contrast, referred instead to just one developmental resource (i.e., the molecular sequence) involved in the complex development of the disease, which interacted with a host of other such resources (proteins, RNA, a variety of enzymes, etc.); Gene-D was indeterminate with regards to the ultimate phenotypic disease. Moreover, in cases of other diseases where there are different disease alleles at the same locus, a Gene-D perspective would treat these alleles as individual genes, while a Gene-P perspective treats them collectively as "the gene for" the disease.

A second philosophical approach for conceptualizing the gene involved rethinking a single, unified gene concept that captured the molecular-developmental complexities. For example, Eva Neumann-Held claimed that a "process molecular gene concept" (PMG) embraced the complicated developmental intricacies. On her unified view, the term "gene" referred to "the recurring process that leads to the temporally and spatially regulated expression of a particular polypeptide product". Returning to the case of cystic fibrosis, a PMG for an individual without the disease referred to one of a variety of transmembrane ion-channel templates along with all the epigenetic factors, i.e., nongenetic influences on gene expression, involved in the generation of the normal polypeptide product. And so cystic fibrosis arose when a particular stretch of the DNA sequence was missing from this process.

Relatedly, philosophers have also debated the causal distinctiveness of DNA. Consider again the case of cystic fibrosis. A stretch of DNA on chromosome 7 is involved in the process of gene expression, which generates (or fails to generate) the functional product that transports chloride ions. But obviously that final product results from that stretch of DNA as well as all the other developmental resources involved in gene expression, be it in the expression of the functional protein or the dysfunctional one. Thus, a number of authors have argued for a causal parity thesis, wherein all developmental resources involved in the generation of a phenotype such as cystic fibrosis are treated as being on par.

Waters, in reply, has argued that there is something causally distinctive about DNA. Causes are often conceived of as being difference makers, in that a variable (i.e., an entity or activity in a mechanism) can be deemed causal when a change in the value of that variable would counterfactually have led to a different. According to Waters, there are a number of *potential* difference makers in the mechanisms involved in developing or not developing cystic fibrosis; that is, an individual with two normal copies of the gene could still display signs of cystic fibrosis if a manipulation was done to the individual's RNA polymerase (the protein responsible for transcribing DNA to RNA), thereby undermining the functional reading of the stretch of DNA. So RNA polymerase is a difference maker in the development or lack of development of cystic fibrosis, but only a potential difference maker, since variation in RNA polymerase does not play a role in the development or lack of development of cystic fibrosis in natural populations. The stretch of DNA on chromosome 7, however, is an *actual* difference maker. That is, there are actual differences in natural human populations on this stretch of DNA, which lead to actual differences in developing or not developing cystic fibrosis; DNA is causally distinctive, according to Waters, because it is an actual difference maker. Advocates of the parity thesis are thus challenged to identify the other resources (in addition to DNA) that are actual difference makers.

Recently, Paul Griffiths and Karola Stotz have responded to this challenge by offering examples in which, depending on context, regulatory mechanisms can either contribute additional information to the gene products or create gene products for which there is no underlying sequence. Thus, according to Griffiths and Stotz, to assign a causally distinctive role to DNA, as Waters does, is to ignore key aspects of how the gene makes its product.

Common Molecular Biology Techniques

The following list covers some of the more commonly used molecular biology techniques – it is by no means exhaustive:

- Electrophoresis – a process which separates molecules such as DNA or proteins out according to their size, electrophoresis is a mainstay of molecular biology laboratories. While knowing the size of a molecule might not seem like all that much information, it can be used to identify molecules or fragments of molecules and as a check to make sure that we have the correct molecule present.

- Polymerase Chain Reaction (PCR) – a process used to amplify very small amounts of DNA to amounts which can be used in further experiments. It is used as a basic tool in molecular biology to ensure that we have sufficient DNA to carry out further techniques such as genetic modification, however it has wider practical uses such as in forensics (identification using DNA profiling) and disease diagnosis. PCR can also be used to introduce small point mutations into a gene in a process called site-directed mutagenesis.
- Restriction Digest – the process of cutting DNA up into smaller fragments using enzymes which only act at a particular genetic sequence.
- Ligation – the process of joining two pieces of DNA together. Ligation is useful when introducing a new piece of DNA into another genome.
- Blotting – a technique used to specifically identify biomolecules following electrophoresis. The molecule of interest is indicated using either a labeled probe (a complementary strand of nucleic acid) or a labeled antibody raised against a specific protein.
- Cloning – the technique of introducing a new gene into a cell or organism. This can be used to see what effect the expression of that gene has on the organism, to turn the organism into a factory which will produce large quantities of the gene or the protein it codes for, or (within the inclusion of a label) to indicate where the products of that gene are expressed in the organism. Insertion of genetic material into a bacterium is called transformation, while insertion into a eukaryotic cell is called transfection. If a virus is used to introduce this material, the process is called transduction.

Each of these techniques is used in conjunction with other techniques to help scientists solve a particular research question. For example, following using PCR to create large quantities of a particular gene a scientist may ligate a gene for a particular protein into a plasmid vector (a short circular strand of DNA which acts as a carrier), perform a quick restriction digest and electrophoresis to ensure that the gene has been inserted properly, and then use that plasmid to transform a bacterial cell which is used to produce large quantities of the vector. After purification of the vector from the bacteria, it is then used to transfect a mammalian cell in culture. The scientist then uses protein electrophoresis and western blotting to demonstrate the expression of the gene product.

References

- Hoppensteadt, F. (1993) [1975]. Mathematical Theories of Populations: Demographics, Genetics and Epidemics(Reprinted ed.). Philadelphia: SIAM. ISBN 0-89871-017-0
- What-is-theoretical-biology: quora.com, Retrieved 25 May 2018
- Strogatz, S. H. (2001). Nonlinear Dynamics and Chaos: Applications to Physics, Biology, Chemistry, and Engineering. Perseus. ISBN 0-7382-0453-6
- Zoology, Science-48756: britannica.com, Retrieved 14 July 2018
- Botany, science: britannica.com, Retrieved 10 June 2018
- Marine-life-definition-and-examples-2291890: thoughtco.com, Retrieved 25 May 2018
- Molecular-biology: aboutbioscience.org, Retrieved 25 April 2018

Permissions

All chapters in this book are published with permission under the Creative Commons Attribution Share Alike License or equivalent. Every chapter published in this book has been scrutinized by our experts. Their significance has been extensively debated. The topics covered herein carry significant information for a comprehensive understanding. They may even be implemented as practical applications or may be referred to as a beginning point for further studies.

We would like to thank the editorial team for lending their expertise to make the book truly unique. They have played a crucial role in the development of this book. Without their invaluable contributions this book wouldn't have been possible. They have made vital efforts to compile up to date information on the varied aspects of this subject to make this book a valuable addition to the collection of many professionals and students.

This book was conceptualized with the vision of imparting up-to-date and integrated information in this field. To ensure the same, a matchless editorial board was set up. Every individual on the board went through rigorous rounds of assessment to prove their worth. After which they invested a large part of their time researching and compiling the most relevant data for our readers.

The editorial board has been involved in producing this book since its inception. They have spent rigorous hours researching and exploring the diverse topics which have resulted in the successful publishing of this book. They have passed on their knowledge of decades through this book. To expedite this challenging task, the publisher supported the team at every step. A small team of assistant editors was also appointed to further simplify the editing procedure and attain best results for the readers.

Apart from the editorial board, the designing team has also invested a significant amount of their time in understanding the subject and creating the most relevant covers. They scrutinized every image to scout for the most suitable representation of the subject and create an appropriate cover for the book.

The publishing team has been an ardent support to the editorial, designing and production team. Their endless efforts to recruit the best for this project, has resulted in the accomplishment of this book. They are a veteran in the field of academics and their pool of knowledge is as vast as their experience in printing. Their expertise and guidance has proved useful at every step. Their uncompromising quality standards have made this book an exceptional effort. Their encouragement from time to time has been an inspiration for everyone.

The publisher and the editorial board hope that this book will prove to be a valuable piece of knowledge for students, practitioners and scholars across the globe.

Index